Springer Tracts in Advanced Robotics 119

Editors

Prof. Bruno Siciliano
Dipartimento di Ingegneria Elettrica
e Tecnologie dell'Informazione
Università degli Studi di Napoli
Federico II
Via Claudio 21, 80125 Napoli
Italy
E-mail: siciliano@unina.it

Prof. Oussama Khatib
Artificial Intelligence Laboratory
Department of Computer Science
Stanford University
Stanford, CA 94305-9010
USA
E-mail: khatib@cs.stanford.edu

More information about this series at http://www.springer.com/series/5208

Rafael Valencia · Juan Andrade-Cetto

Mapping, Planning and Exploration with Pose SLAM

Springer

Rafael Valencia
Carnegie Mellon University
Pittsburgh, PA
USA

Juan Andrade-Cetto
CSIC-UPC
Institut de Robòtica i Informàtica Industrial
Barcelona
Spain

ISSN 1610-7438 ISSN 1610-742X (electronic)
Springer Tracts in Advanced Robotics
ISBN 978-3-319-60602-6 ISBN 978-3-319-60603-3 (eBook)
DOI 10.1007/978-3-319-60603-3

Library of Congress Control Number: 2017943233

Printed on acid-free paper

This Springer imprint is published by Springer Nature
The registered company is Springer International Publishing AG
The registered company address is: Gewerbestrasse 11, 6330 Cham, Switzerland

To Graciela, Silvia and M. Rafael.

Foreword

Robotics is undergoing a major transformation in scope and dimension. From a largely dominant industrial focus, robotics is rapidly expanding into human environments and vigorously engaged in its new challenges. Interacting with, assisting, serving, and exploring with humans, the emerging robots will increasingly touch people and their lives.

Beyond its impact on physical robots, the body of knowledge robotics has produced is revealing a much wider range of applications reaching across diverse research areas and scientific disciplines, such as biomechanics, haptics, neurosciences, virtual simulation, animation, surgery, and sensor networks among others. In return, the challenges of the new emerging areas are proving an abundant source of stimulation and insights for the field of robotics. It is indeed at the intersection of disciplines that the most striking advances happen.

The Springer Tracts in Advanced Robotics (STAR) is devoted to bringing to the research community the latest advances in the robotics field on the basis of their significance and quality. Through a wide and timely dissemination of critical research developments in robotics, our objective with this series is to promote more exchanges and collaborations among the researchers in the community and contribute to further advancements in this rapidly growing field.

The monograph by R. Valencia and J. Andrade Cetto is based on the first author's Ph.D. thesis work. It deals with the mapping, path planning, and autonomous exploration problems, adopting the so-called Pose SLAM as the basic state estimation machinery. A novel approach allowing a mobile robot to plan a path and to select the appropriate actions to autonomously construct the map is proposed, while maximizing coverage and minimizing localization and map uncertainties. The method has been extensively tested both in simulation and in experiments with a real outdoor robot.

STAR is proud to welcome yet another volume in the series dedicated to the popular area of SLAM!

Naples, Italy
February 2017

Bruno Siciliano
STAR Editor

Preface

This work reports research on mapping, path planning, and autonomous exploration. These are classical problems in robotics, typically studied independently, and here we link such problems by framing them within a common SLAM approach, adopting Pose SLAM as the basic state estimation machinery. The main contribution of this work is an approach that allows a mobile robot to plan a path using the map it builds with Pose SLAM and to select the appropriate actions to autonomously construct this map.

Pose SLAM is the variant of SLAM where only the robot trajectory is estimated and where landmarks are only used to produce relative constraints between robot poses. In Pose SLAM, observations come in the form of relative-motion measurements between robot poses. With regards to extending the original Pose SLAM formulation, this work studies the computation of such measurements when they are obtained with stereo cameras and develops the appropriate noise propagation models for such case. Furthermore, the initial formulation of Pose SLAM assumes poses in SE(2) and in this work we extend this formulation to SE(3), parameterizing rotations either with Euler angles and quaternions. We also introduce a loop closure test that exploits the information from the filter using an independent measure of information content between poses. In the application domain, we present a technique to process the 3D volumetric maps obtained with this SLAM methodology, but with laser range scanning as the sensor modality, to derive traversability maps that were useful for the navigation of a heterogeneous fleet of mobile robots in the context of the EU project URUS.

Aside from these extensions to Pose SLAM, the core contribution of the work is an approach for path planning that exploits the modeled uncertainties in Pose SLAM to search for the path in the pose graph with the lowest accumulated robot pose uncertainty, i.e., the path that allows the robot to navigate to a given goal with the least probability of becoming lost. An added advantage of the proposed path planning approach is that since Pose SLAM is agnostic with respect to the sensor modalities used, it can be used in different environments and with different robots, and since the original pose graph may come from a previous mapping session, the paths stored in the map already satisfy constraints not easy modeled in the robot controller, such as the existence of restricted regions, or the right of way along paths. The proposed path planning methodology has been extensively tested both in simulation and with a real outdoor robot.

Our path planning approach is adequate for scenarios where a robot is initially guided during map construction, but autonomous during execution. For other scenarios in which more autonomy is required, the robot should be able to explore the environment without any supervision. The second core contribution of this work is an autonomous exploration method that complements the aforementioned path planning strategy. The method selects the appropriate actions to drive the robot so as to maximize coverage and at the same time minimize localization and map uncertainties. An occupancy grid is maintained for the sole purpose of guaranteeing coverage. A significant advantage of the method is that since the grid is only computed to hypothesize entropy reduction of candidate map posteriors, it can be computed at a very coarse resolution since it is not used to maintain neither the robot localization estimate, nor the structure of the environment. Our technique evaluates two types of actions: exploratory actions and place revisiting actions. Action decisions are made based on entropy reduction estimates. By maintaining a Pose SLAM estimate at run time, the technique allows to replan trajectories online should significant change in the Pose SLAM estimate be detected. The proposed exploration strategy was tested in a common publicly available dataset comparing favorably against frontier based exploration.

This work was supported in part by a Ph.D. scholarship from the Mexican Council of Science and Technology (CONACYT), a Research Stay Grant from the Agència de Gestió d'Ajuts Universitaris i de Recerca of The Generalitat of Catalonia (2010 BE1 00967), the Consolidated Research Group VIS (SGR2009-2013), a Technology Transfer Contract with the Asociación de la Industria Navarra, the European Integrated Project ARCAS (FP7-ICT-287617), the National Research Projects PAU (DPI2008-06022), PAU+ (DPI2011-2751) and RobInstruct: Instructing robots using natural communication skills TIN2014-58178-R funded by the Spanish Ministry of Economy and Competitiveness.

Pittsburgh, USA
Barcelona, Spain
February 2017

Rafael Valencia
Juan Andrade-Cetto

Contents

Chapter 1
Introduction

Simultaneous localization and mapping (SLAM) is the process where a mobile robot builds a map of an unknown environment while at the same time being localized relative to this map. Performing SLAM is a basic task for a truly autonomous robot. Consequently, it has been one of the main research topics in robotics for the last two decades. Whereas in the seminal approaches to SLAM [1] only few tens of landmarks could be managed, state of the art approaches can now efficiently manage thousands of landmarks [2–4] and build maps over several kilometers [5].

Despite these important achievements in SLAM research, very little has been investigated concerning approaches that allow the robot to actually employ the maps it builds for navigation. Aside from applications such as the reconstruction of archaeological sites [6] or the inspection of dangerous areas [7], the final objective for an autonomous robot is not to build a map of the environment, but to use this map for navigation. Another issue that has not received extensive attention is the problem of autonomous exploration for SLAM. Most SLAM techniques are passive in the sense that the robot only estimates the model of the environment, but without taking any decisions on its trajectory.

The work reported by this book corresponds to Rafael Valencia's doctoral thesis [8] directed by Juan Andrade-Cetto. This work mainly contributes with an approach that allows a mobile robot to plan a path using the map it builds with SLAM and to select the appropriate actions to autonomously construct this map. In addition, it studies related issues such as visual odometry and 3D mapping. Thus, this work reports research on mapping, path planning, and autonomous exploration. These are classical problems in robotics, typically studied independently, and here we link such problems by framing them within a common SLAM approach.

In this work we adopt the Pose SLAM approach [9] as the basic state estimation machinery. Pose SLAM is the variant of SLAM where only the robot trajectory is estimated and where landmarks are only used to produce relative constraints between robot poses. Thus, the map in Pose SLAM only contains the trajectory of the robot.

R. Valencia and J. Andrade-Cetto, *Mapping, Planning and Exploration with Pose SLAM*, Springer Tracts in Advanced Robotics 119,
DOI 10.1007/978-3-319-60603-3_1

The poses stored in the map are, by construction, feasible and obstacle-free since they were already traversed by the robot when the map was originally built. Additionally, Pose SLAM only keeps non-redundant poses and highly informative links. Thus, the state does not grow independently of the size of the environment. It also translates into a significant reduction of the computational cost and a delay of the filter inconsistency, maintaining the quality of the estimation for longer mapping sequences.

In Pose SLAM, observations come in the form of relative-motion measurements between any two robot poses. This work studies the computation of such measurements when they are obtained with stereo cameras and presents an implementation of a visual odometry method that includes a noise propagation technique.

The initial formulation of Pose SLAM [9] assumes poses in SE(2) and in this work we extend this formulation to poses in SE(3), parameterizing rotations either with Euler angles and quaternions. We also introduce a loop closure test tailored to Pose SLAM that exploits the information from the filter using an independent measure of information content between poses, which for consistent estimates is less affected by perceptual aliasing. Furthermore, we present a technique to process the 3D volumetric maps obtained with this SLAM implementation in SE(3) to derive traversability maps useful for the navigation of a heterogeneous fleet of mobile robots.

Besides the aforementioned advantages of Pose SLAM, a notable property of this approach for the purposes of this work is that, unlike standard feature-based SLAM, its map can be directly used for path planning. The reason that feature-based SLAM cannot be directly used to plan trajectories is that these methods produce a sparse graph of landmark estimates and their probabilistic relations, which is of little value to find collision free paths for navigation. These graphs can be enriched with obstacle related information [10–12], but it increases the complexity. On the contrary, as the outcome of Pose SLAM is a graph of obstacle-free paths in the area where the robot has been operated, this map can be directly employed for path planning.

In this work we propose an approach for path planning under uncertainty that exploits the modeled uncertainties in robot poses by Pose SLAM to search for the path in the pose graph with the lowest accumulated robot pose uncertainty, i.e., the path that allow the robot to navigate to the goal without becoming lost.

The approach from the motion planning literature that best matches our path planning approach is the Belief Roadmap (BRM) [13, 14]. In such an approach, the edges defining the roadmap include information about the uncertainty change when traversing such an edge. However, the main drawback of the BRM is that it still assumes a known model of the environment, which is in general not available in real applications. In contrast, we argue in this work that Pose SLAM graphs can be directly used as belief roadmaps.

An added advantage of our path planning approach is that Pose SLAM is agnostic with respect to the sensor modalities used, which facilitates its application in different environments and robots, and the paths stored in the map satisfy constraints not easy to model in the robot controller, such as the existence of restricted regions, or the right of way along paths.

Our path planning approach is adequate for scenarios where a robot is initially guided during map construction, but autonomous during execution. For other scenarios in which more autonomy is required, the robot should be able to explore the environment without any supervision. In this work we also introduce an autonomous exploration approach for the case of Pose SLAM, which complements the path planning method.

A straightforward solution to the problem of exploration for SLAM is to combine a classical exploration method with SLAM. However, classical exploration methods focus on reducing the amount of unseen area disregarding the cumulative effect of localization drift, leading the robot to accumulate more and more uncertainty. Thus, a solution to this problem should revisit known areas from time to time, trading off coverage with accuracy.

In this work we propose an autonomous exploration strategy for the case of Pose SLAM that automates the belief roadmap construction from scratch by selecting the appropriate actions to drive the robot so as to maximize coverage and at the same time minimize localization and map uncertainties. In our approach, we guarantee coverage with an occupancy grid of the environment. A significant advantage of the approach is that this grid is only computed to hypothesize entropy reduction of candidate map posteriors, and that it can be computed at a very coarse resolution since it is not used to maintain neither the robot localization estimate, nor the structure of the environment. In a similar way to [15], our technique evaluates two types of actions: exploratory actions and place revisiting actions. Action decisions are made based on entropy reduction estimates. By maintaining a Pose SLAM estimate at run time, the technique allows to replan trajectories online should significant change in the Pose SLAM estimate be detected, something that would make the computed entropy reduction estimates obsolete.

This work is structured in three parts. The first is devoted to the SLAM approach we employ along the work. The second part introduces our algorithm for planning under uncertainty. Finally, in the last part we present an exploration approach to automate the map building process with Pose SLAM. Figure 1.1 shows a block diagram representing the system architecture proposed in this work that also outlines the structure of this document.

In this work we follow the abstraction of SLAM usually employed by Pose graph SLAM methods [16, 17], which divides SLAM in a front-end and a back-end part. Thus, the first part of the work is split in two chapters. We begin our discussion by presenting our SLAM front-end in Chap. 2, which is in charge of processing the sensor information to compute the relative-motion measurements. In the context of this work observations come in the form of relative-motion constraints between two robot poses. These are typically computed using the Iterative Closest Point (ICP) method [18] when working with laser scans. When working with stereo images, visual odometry techniques are usually employed to recover the relative-pose measurements. The latter method is adopted in our contribution presented at IROS 2007 [19] and in Chap. 2 we describe it in more detail and introduce a technique to model the measurements noise, which propagates the noise in image features to relative-pose measurements.

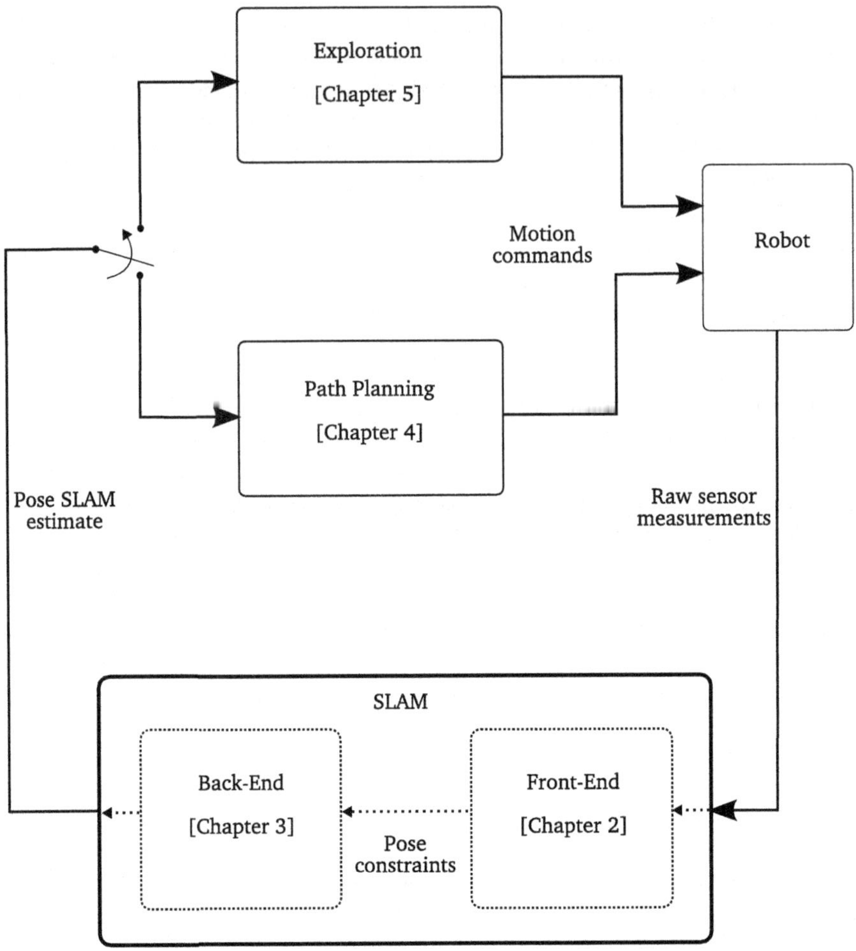

Fig. 1.1 System architecture and work outline

Next, in Chap. 3, we present the back-end part of Pose SLAM, that is, the related to the state estimation task, which is sensor agnostic. We begin with an exposition of the basics of Pose SLAM based on the work by Eustice et al. [20] and Ila et al. [9]. In this exposition we also include one of our initial contributions, consisting of a loop closure test for Pose SLAM, presented at IROS 2007 [19]. Next, we discuss the extension of Pose SLAM to deal with poses in SE(3) and show results on its application to build 3D volumetric maps and traversability maps. Such results were presented at IROS 2009 [21] and were developed as part of the European Union-funded project "Ubiquitous networking robotics in urban settings" (URUS) [22]. Furthermore, the maps we built were also employed for the calibration of a camera network tailored for this project, whose results were presented at the Workshop on Network Robot Systems at IROS 2009 [23].

The second part of this work deals with the problem of path planning with SLAM. Chapter 4 details our path planning method. It describes how to plan a path using the roadmap built with Pose SLAM, presents the new planning approach, and shows results with datasets and real world experiments with a four-wheel mobile robot. An initial version of this approach was documented in a technical report [24] and, eventually, an improved version was presented at ICRA 2011 [25]. A journal version of this work that includes more improvements as well as real world experiments appeared at IEEE Transactions on Robotics [26]. Furthermore, results on path planning with 3D volumetric maps appeared at the Spanish workshop ROBOT'11 [27].

Lastly, our autonomous exploration strategy for Pose SLAM is presented in Chap. 5. This part of the work was carried out during a stay at the Centre for Autonomous Systems, in the Faculty of Engineering and Information Technology, at the University of Technology, Sydney. The results of this approach were presented at IROS 2012 [28].

References

1. R. Smith, M. Self, P. Cheeseman, A stochastic map for uncertain spatial relationships, in *Proceedings of 4th International Symposium Robotics Research* (Santa Clara, 1988), pp. 467–474
2. K. Konolige, Large-scale map-making, in *Proceedings of 19th AAAI Conference on Artificial Intelligence* (San Jose, California, Jul 2004) pp. 457–463
3. S. Thrun, M. Montemerlo, The graph SLAM algorithm with applications to large-scale mapping of urban structures. Int. J. Robot. Res. **25**(5–6), 403–429 (2006)
4. U. Frese, L. Schröder, Closing a million-landmarks loop, in *Proceedings of IEEE/RSJ International Conference on Intelligent Robots and Systems* (Beijing, Oct 2006), pp. 5032–5039
5. G. Sibley, C. Mei, I. Reid, P. Newman, Vast-scale outdoor navigation using adaptive relative bundle adjustment. Int. J. Robot. Res. **29**(8), 958–980 (2010)
6. R.M. Eustice, H. Singh, J.J. Leonard, M.R. Walter, Visually mapping the RMS Titanic: conservative covariance estimates for SLAM information filters. Int. J. Robot. Res. **25**(12), 1223–1242 (2006)
7. S. Thrun, S. Thayer, W. Whittaker, C. Baker, W. Burgard, D. Ferguson, D. Hahnel, M. Montemerlo, A. Morris, Z. Omohundro, C. Reverte, Autonomous exploration and mapping of abandoned mines. Robot. Automat. Mag. **11**(4), 79–91 (2004)
8. R. Valencia, Mapping, planning and exploration with Pose SLAM. Ph.D. thesis, UPC, Barcelona, Apr 2013, http://hdl.handle.net/10803/117471
9. V. Ila, J.M. Porta, J. Andrade-Cetto, Information-based compact Pose SLAM. IEEE Trans. Robot. **26**(1), 78–93 (2010)
10. P.M. Newman, J. Leonard, J. Neira, J.D. Tardós, Explore and return: Experimental validation of real time concurrent mapping and localization, in *Proceedings of IEEE International Conference on Robotics and Automation* (Washington, May 2002), pp. 1802–1809
11. J. Guivant, E. Nebot, J. Nieto, F. Masson, Navigation and mapping in large unstructured environments. Int. J. Robot. Res. **23**(4–5), 449–472 (2004)
12. S. Rezaei, J. Guivant, J. Nieto, E. Nebot, Simultaneous information and global motion analysis ("SIGMA") for car-like robots, in *Proceedings of IEEE International Conference on Robotics and Automation* (New Orleans, Apr 2004), pp 1939–1944

13. S. Prentice, N. Roy, The belief roadmap: efficient planning in belief space by factoring the covariance. Int. J. Robot. Res. **29**(11–12), 1448–1465 (2009)
14. R. He, S. Prentice, N. Roy, Planning in information space for a quadrotor helicopter in a GPS-denied environment, in *Proceedings of IEEE International Conference on Robotics and Automation* (Pasadena, May 2008), pp. 1814–1820
15. C. Stachniss, G. Grisetti, W. Burgard, Information gain-based exploration using Rao-Blackwellized particle filters, in *Robotics: Science and Systems I* (Cambridge, Jun 2005), pp. 65–72
16. E. Olson, J. Leonard, S. Teller, Fast iterative alignment of pose graphs with poor initial estimates, in *Proceedings of IEEE International Conference on Robotics and Automation* (Orlando, May 2006), pp. 2262–2269
17. G. Grisetti, C. Stachniss, S. Grzonka, W. Burgard, A tree parameterization for efficiently computing maximum likelihood maps using gradient descent, in *Robotics: Science and Systems III* (Atlanta, Jun 2007), pp. 9:1–9:8
18. P.J. Besl, N.D. McKay, A method for registration of 3D shapes. IEEE Trans. Pattern Anal. Mach. Intell. **14**(2), 239–256 (1992)
19. V. Ila, J. Andrade-Cetto, R. Valencia, A. Sanfeliu, Vision-based loop closing for delayed state robot mapping, in *Proceedings of IEEE/RSJ International Conference on Intelligent Robots and Systems* (San Diego, Nov 2007), pp. 3892–3897
20. R.M. Eustice, H. Singh, J.J. Leonard, Exactly sparse delayed-state filters for view-based SLAM. IEEE Trans. Robot. **22**(6), 1100–1114 (2006)
21. R. Valencia, E.H. Teniente, E. Trulls, J. Andrade-Cetto, 3D mapping for urban service robots, in *Proceedings of IEEE/RSJ International Conference on Intelligent Robots and Systems* (Saint Louis, Oct 2009), pp. 3076–3081
22. A. Sanfeliu, J. Andrade-Cetto, Ubiquitous networking robotics in urban settings, in *Proceedings of IEEE/RSJ IROS Workshop on Network Robot Systems* (Beijing, Oct 2006), pp. 14–18
23. J. Andrade-Cetto, A. Ortega, E. Teniente, E. Trulls, R. Valencia, A. Sanfeliu, Combination of distributed camera network and laser-based 3d mapping for urban service robotics, in *Proceedings of IEEE/RSJ IROS Workshop on Network Robot Systems* (Saint Louis, Oct 2009), pp. 69–80
24. R. Valencia, J. Andrade-Cetto, J.M Porta, Path planning with Pose SLAM. Technical Report IRI-DT-10-03, IRI, UPC, 2010
25. R. Valencia, J. Andrade-Cetto, J. M. Porta, Path planning in belief space with Pose SLAM, in *Proceedings of IEEE International Conference on Robotics and Automation* (Shanghai, May 2011), pp. 78–83
26. R. Valencia, M. Morta, J. Andrade-Cetto, J.M. Porta, Planning reliable paths with Pose SLAM. IEEE Trans. Robot. **29**(4), 1050–1059 (2013)
27. E.H. Teniente, R. Valencia, J. Andrade-Cetto, Dense outdoor 3D mapping and navigation with Pose SLAM, in *Proceedings of III Workshop de Robótica: Robótica Experimental, ROBOT'11* (Seville, 2011), pp. 567–572
28. R. Valencia, J. Valls Miró, G. Dissanayake, J. Andrade-Cetto, Active Pose SLAM, in *Proceedings of IEEE/RSJ International Conference on Intelligent Robots and Systems* (Algarve, Oct 2012), pp. 1885–1891

Chapter 2
SLAM Front-End

In this Chapter we discuss our choice of front-end for SLAM, the part in charge of processing the sensor information to generate the observations that will be fed to the estimation machinery. In the context of this work, observations come in the form of relative-motion constraints between any two robot poses. They are typically obtained with the Iterative Closest Point (ICP) algorithm [1] when working with laser scans. When using stereo images, the egomotion of the robot can be estimated with visual odometry [2, 3]. The latter method is adopted in our contribution presented in [4] and in this Chapter we describe it in more detail and extend it with a technique to model the uncertainty of the relative-motion constraints.

Assuming we have a pair of stereo images acquired with two calibrated cameras fixed to the robot's frame, our approach iterates as follows: SIFT image features [5] are extracted from the four images and matched between them. The resulting point correspondences are used for least-squares stereo reconstruction. Next, matching of these 3D features in the two consecutive frames is used to compute a least-squares best-fit pose transformation, rejecting outliers via RANSAC [6].

However, the outcome of this approach is also prone to errors. Errors in locating the image features lead to errors in the location of the 3D feature points after stereo reconstruction, which eventually cause errors in the motion estimate. Modeling such error propagation allows to compute motion estimates with the appropriate uncertainty bounds. In this Chapter we introduce a technique to compute the covariance of the relative pose measurement by first-order error propagation [7].

These camera pose constraints are eventually used as relative pose measurements in the SLAM we employ in this work. They are used either as odometry measurements, when matching stereo images from consecutive poses in time, or as loop closure constraints, when computing the relative motion of the last pose with respect to any previous pose. This will be discussed in Chap. 3.

R. Valencia and J. Andrade-Cetto, *Mapping, Planning and Exploration with Pose SLAM*, Springer Tracts in Advanced Robotics 119,
DOI 10.1007/978-3-319-60603-3_2

The rest of this Chapter is structured as follows. Section 2.1 explains the feature extraction and the stereo reconstruction process. Next, the pose estimation step is shown in Sect. 2.2. Then, in Sect. 2.3 we introduce a technique to model the uncertainty of the relative motion measurement. Finally, in Sect. 2.4 we provide bibliographical notes.

2.1 Feature Extraction and Stereo Reconstruction

Simple correlation-based features, such as Harris corners [8] or Shi and Tomasi features [9], are of common use in vision-based SFM and SLAM; from the early uses of Harris himself to the popular work of Davison [10]. This kind of features can be robustly tracked when camera displacement is small and are tailored to real-time applications. However, given their sensitivity to scale, their matching is prone to fail under larger camera motions; less to say for loop-closing hypotheses testing. Given their scale and local affine invariance properties, we opt to use SIFTs instead [5, 11], as they constitute a better option for matching visual features from significantly different vantage points.

In our system, features are extracted and matched with previous image pairs. Then, from the surviving features, we compute the imaged 3D scene points as follows.

Assumming two stereo-calibrated cameras and a pin-hole camera model [12], with the left camera as the reference of the stereo system, the following expressions relate a 3D scene point $\mathbf{p}$ to the corresponding points $\mathbf{m} = [u, v]^\top$ in the left, and $\mathbf{m}' = \left[u', v'\right]^\top$ in the right camera image planes

$$\begin{bmatrix} \mathbf{m} \\ s \end{bmatrix} = \begin{bmatrix} \alpha_u & 0 & u_o & 0 \\ 0 & \alpha_v & v_o & 0 \\ 0 & 0 & 1 & 0 \end{bmatrix} \begin{bmatrix} \mathbf{I}_3 & 0 \\ \mathbf{0}_{1\times 3} & 1 \end{bmatrix} \begin{bmatrix} \mathbf{p} \\ 1 \end{bmatrix}, \tag{2.1}$$

$$\begin{bmatrix} \mathbf{m}' \\ s' \end{bmatrix} = \begin{bmatrix} \alpha'_u & 0 & u'_o & 0 \\ 0 & \alpha'_v & v'_o & 0 \\ 0 & 0 & 1 & 0 \end{bmatrix} \begin{bmatrix} \mathbf{R} & \mathbf{t} \\ \mathbf{0}_{1\times 3} & 1 \end{bmatrix} \begin{bmatrix} \mathbf{p} \\ 1 \end{bmatrix}, \tag{2.2}$$

where α_u and α_v are the pixel focal lengths in the x and y directions for the left camera, and α'_u and α'_v for the right camera, (u_o, v_o) and (u'_o, v'_o) are the left and right camera image centers, respectively. The homogeneous transformation from the right camera frame to the reference frame of the stereo system is represented by the rotation matrix $\mathbf{R}$ and translation vector $\mathbf{t} = [t_x, t_y, t_z]^\top$. $[\mathbf{m}^\top, s]^\top$ and $[\mathbf{m}'^\top, s']^\top$ are the left and right image points in homogeneous coordinates, with scale s and s', respectively, and $\mathbf{I}_3$ is a 3×3 identity matrix.

Equations 2.1 and 2.2 define the following overdetermined system of equations

$$\begin{bmatrix} (u' - u'_o)\mathbf{r}_3^\top - \alpha'_u \mathbf{r}_1^\top \\ (v' - v'_o)\mathbf{r}_3^\top - \alpha'_v \mathbf{r}_2^\top \\ -\alpha_u, 0, u - u_o \\ 0, -\alpha_v, v - v_o \end{bmatrix} \begin{bmatrix} x \\ y \\ z \end{bmatrix} = \begin{bmatrix} (u'_o - u')t_z + \alpha'_u t_x \\ (v'_o - v')t_z + \alpha'_v t_y \\ 0 \\ 0 \end{bmatrix}$$

$$\mathbf{A}\,\mathbf{p} = \mathbf{b}, \tag{2.3}$$

where $\mathbf{R}$ is expressed by its row elements

$$\mathbf{R} = \begin{bmatrix} \mathbf{r}_1^\top \\ \mathbf{r}_2^\top \\ \mathbf{r}_3^\top \end{bmatrix}.$$

Solving for $\mathbf{p}$ in Eq. 2.3 gives the sought 3D coordinates of the imaged points $\mathbf{m}$ and $\mathbf{m}'$. Performing this process for each pair of matching feature in a pair of stereo images results in two sets of 3D points, or 3D point clouds, i.e. $\left\{\mathbf{p}_1^{(i)}\right\}$ and $\left\{\mathbf{p}_2^{(i)}\right\}$.

2.2 Pose Estimation

Next, we present two alternatives to compute the relative motion of the camera from two stereo images by solving the 3D to 3D pose estimation problem.

The general solution to this problem consists of finding the rotation matrix $\mathbf{R}$ and translation vector $\mathbf{t}$ that minimize the squared L_2-norm for all points in the two aforementioned clouds,

$$\left\{\hat{\mathbf{R}}, \hat{\mathbf{t}}\right\} = \underset{\mathbf{R},\mathbf{t}}{\operatorname{argmin}} \sum_{i=1}^{N} \left\| \mathbf{p}_1^{(i)} - \left(\mathbf{R}\mathbf{p}_2^{(i)} + \mathbf{t}\right) \right\|^2, \tag{2.4}$$

with N the number of points in each cloud.

For both methods, we resort to the use of RANSAC [6] to eliminate outliers. It might be the case that SIFT matches occur on areas of the scene that experienced motion during the acquisition of the two image stereo pairs. For example, an interest point might appear at an acute angle of a tree leaf shadow, or on a person walking in front of the robot. The corresponding matching 3D points will not represent good fits to the camera motion model, and might introduce large bias to our least squares pose error minimization. The use of such a robust model fitting technique allows us to preserve the largest number of point matches that at the same time minimize the square sum of the residuals, as shown in Fig. 2.1.

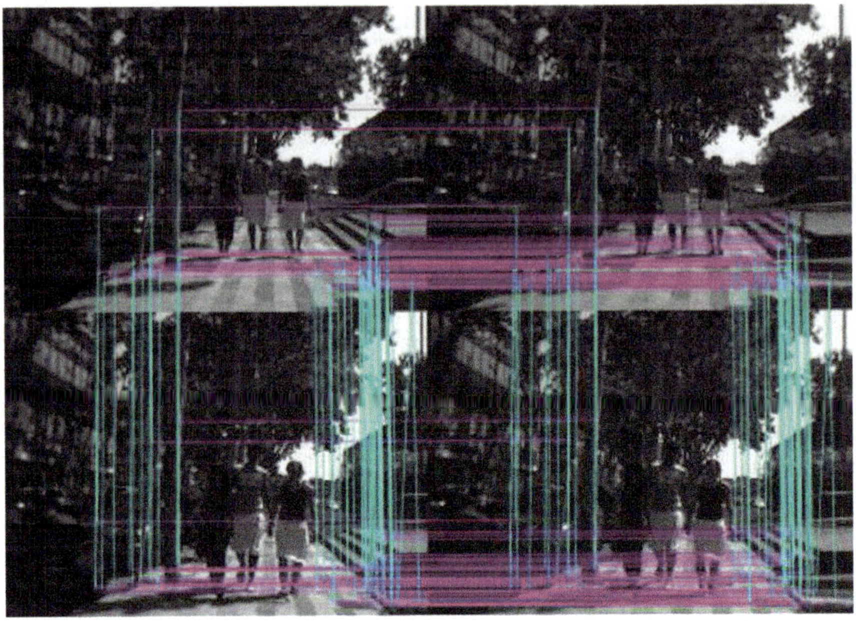

Fig. 2.1 SIFT correspondences in two consecutive stereo image pairs after outlier removal using RANSAC

Furthermore, if the covariance of the matching points is available it can be exploited so as to explicitly model their precision according to their distance from the camera. For instance, we can weight the point mismatch in Eq. 2.4 with the covariance of the triangulation of the two points. However, this would complicate further the optimization problem defined by Eq. 2.4. Instead, we chose to rely on standard techniques such as the following solutions.

2.2.1 *Horn's Method*

A solution for the rotation matrix $\mathbf{R}$ is computed by minimizing the sum of the squared errors between the rotated directional vectors of feature matches for the two robot poses [13]. Directional vectors $\boldsymbol{\nu}$ are computed as the unit norm direction along the imaged 3D scene point $\mathbf{p}$ and indicates the orientation of such a point, that is,

$$\boldsymbol{\nu}_1^{(i)} = \frac{\mathbf{p}_1^{(i)}}{\|\mathbf{p}_1^{(i)}\|} \tag{2.5}$$

and

$$\boldsymbol{v}_2^{(i)} = \frac{\mathbf{p}_2^{(i)}}{\|\mathbf{p}_2^{(i)}\|} \tag{2.6}$$

are the directional vectors for the i-th point on the first and the second point cloud, respectively.

The solution to this minimization problem gives an estimate of the orientation of one cloud of points with respect to the other, and can be expressed in quaternion form as

$$\frac{\partial}{\partial \mathbf{R}} \left(\mathbf{q}^\top \mathbf{B}\mathbf{q}\right) = 0\,, \tag{2.7}$$

where $\mathbf{B}$ is given by

$$\mathbf{B} = \sum_{i=1}^{N} \mathbf{B}_i \mathbf{B}_i^\top\,, \tag{2.8}$$

$$\mathbf{B}_i = \begin{bmatrix} 0 & -c_x^{(i)} & -c_y^{(i)} & -c_z^{(i)} \\ c_x^{(i)} & 0 & b_z^{(i)} & -b_y^{(i)} \\ c_y^{(i)} & -b_z^{(i)} & 0 & b_x^{(i)} \\ c_z^{(i)} & b_y^{(i)} & -b_x^{(i)} & 0 \end{bmatrix}, \tag{2.9}$$

and

$$\mathbf{b}^{(i)} = \boldsymbol{v}_2^{(i)} + \boldsymbol{v}_1^{(i)}, \quad \mathbf{c}^{(i)} = \boldsymbol{v}_2^{(i)} - \boldsymbol{v}_1^{(i)} \quad . \tag{2.10}$$

The quaternion $\mathbf{q}$ that minimizes the argument of the derivative operator in Eq. 2.7 is the smallest eigenvector of the matrix $\mathbf{B}$.

If we denote this smallest eigenvector by the 4-tuple $(q_1, q_2, q_3, q_4)^\top$, it follows that the angle θ associated with the rotational transform is given by

$$\theta = 2cos^{-1}(q_4), \tag{2.11}$$

and the axis of rotation would be given by

$$\hat{\mathbf{a}} = \frac{(q_1, q_2, q_3)^\top}{\sin(\theta/2)}. \tag{2.12}$$

Then, it can be shown that the elements of the rotation submatrix $\mathbf{R}$ are related to the orientation parameters $\hat{\mathbf{a}}$ and θ by

$$\hat{\mathbf{R}} = \begin{bmatrix} a_x^2 + (1 - a_x^2)c_\theta & a_x a_y c'_\theta - a_z s_\theta & a_x a_z c'_\theta + a_y s_\theta \\ a_x a_y c'_\theta + a_z s_\theta & a_y^2 + (1 - a_y^2)c_\theta & a_y a_z c'_\theta - a_x s_\theta \\ a_x a_z c'_\theta - a_y s_\theta & a_y a_z c'_\theta + a_x s_\theta & a_z^2 + (1 - a_z^2)c_\theta \end{bmatrix}, \tag{2.13}$$

where $s_\theta = \sin\theta$, $c_\theta = \cos\theta$, and $c'_\theta = 1 - \cos\theta$.

Once the rotation matrix $\hat{\mathbf{R}}$ is computed, we can use again the matching set of points to compute the translation vector $\hat{\mathbf{t}}$

$$\hat{\mathbf{t}} = \frac{1}{N}\left(\sum_{i=1}^{N} \mathbf{p}_1^{(i)} - \mathbf{R}\sum_{i=1}^{N} \mathbf{p}_2^{(i)}\right). \tag{2.14}$$

2.2.2 SVD-based Solution

This solution decouples the translational and rotation parts of the pose estimation problem by noting that, at the least-squares solution to Eq. 2.4, both of the two 3D point clouds should have the same centroid [14].

Thus, the rotation matrix is computed first by reducing the original least-squares problem to finding the rotation that minimizes

$$\sum_{i=1}^{N} \left\| \bar{\mathbf{p}}_1^{(i)} - \left(\mathbf{R}\bar{\mathbf{p}}_2^{(i)}\right) \right\|^2, \tag{2.15}$$

where

$$\bar{\mathbf{p}}_1^{(i)} = \mathbf{p}_1^{(i)} - \mathbf{c}_1 \tag{2.16}$$

and

$$\bar{\mathbf{p}}_2^{(i)} = \mathbf{p}_2^{(i)} - \mathbf{c}_2, \tag{2.17}$$

express the i-th point on the two point clouds translated to their corresponding centroids, with $\mathbf{c}_1$ and $\mathbf{c}_2$ the centroids of the first and the second point cloud, respectively.

In order to minimize Eq. 2.15, it is defined the 3×3 matrix $\mathbf{M}$

$$\mathbf{M} = \sum_{i=0}^{N} \bar{\mathbf{p}}_1^{(i)} \; \bar{\mathbf{p}}_2^{(i)\top}, \tag{2.18}$$

where its singular value decomposition is given by

$$\mathbf{M} = \mathbf{U}\,\mathbf{\Sigma}\,\mathbf{V}^\top. \tag{2.19}$$

With this, the rotation matrix that minimizes Eq. 2.15 is

$$\hat{\mathbf{R}} = \mathbf{U}\,\mathbf{V}^\top \tag{2.20}$$

as long as $|\mathbf{U}\,\mathbf{V}^\top| = +1$. Otherwise, if $|\mathbf{U}\,\mathbf{V}^\top| = -1$, the solution is a reflection.

Finally, having found the rotation $\hat{\mathbf{R}}$, the translation is computed by

$$\hat{\mathbf{t}} = \mathbf{c}_1 - \hat{\mathbf{R}}\mathbf{c}_2. \tag{2.21}$$

2.3 Error Propagation

In this section we present a method to model the uncertainty of the relative motion measurements computed with the visual odometry approach just described in this Chapter. This method propagates the noise from each matching feature, along the visual odometry process, to end up with a relative pose covariance estimate.

One way to do this is by Monte Carlo simulation, however, this process is time-consuming. Instead, we opt for a closed-form computation based on first order error propagation. That is, given a continuously differentiable function $\mathbf{y} = \mathbf{f}(\mathbf{x})$ and the covariance $\mathbf{\Sigma}_x$ of the input $\mathbf{x}$, we can obtain the covariance $\mathbf{\Sigma}_y$ of the output $\mathbf{y}$ by linearizing $\mathbf{f}(\mathbf{x})$ around the expected value $\mathbf{x_o}$ by a first-order Taylor series expansion. Thus, the first-order error propagation to covariance $\mathbf{\Sigma}_y$ is given by

$$\mathbf{\Sigma}_y = \nabla\mathbf{f}\mathbf{\Sigma}_x\nabla\mathbf{f}^\top,$$

where $\nabla\mathbf{f}$ is the Jacobian of $\mathbf{f}$.

However, sometimes we might not have access to an explicit expression for $\mathbf{y} = \mathbf{f}(\mathbf{x})$, as it will be shown to be our case. Fortunately, though, we still can compute an expression for the Jacobian of $\mathbf{f}(\mathbf{x})$ by the implicit function theorem, which we introduce next.

The implicit function theorem can be stated as follows [7]:

Let $S \subset \mathbb{R}^n \times \mathbb{R}^m$ be an open set and let $\mathbf{\Phi} : S \to \mathbb{R}^m$ be a differentiable function. Suppose that $(\mathbf{x_o}, \mathbf{y_o}) \in S$ that $\mathbf{\Phi}(\mathbf{x_o}, \mathbf{y_o}) = \mathbf{0}$, and that $\left|\frac{\partial\mathbf{\Phi}}{\partial\mathbf{y}}\right|_{(\mathbf{x_o},\mathbf{y_o})} \neq \mathbf{0}$. Then there is an open neighborhood $X \subset \mathbb{R}^n$ of $\mathbf{x_o}$, a neighborhood $Y \subset \mathbb{R}^m$ of $\mathbf{y_o}$, and a unique differentiable function $\mathbf{f} : X \to Y$ such that

$$\mathbf{\Phi}(\mathbf{x}, \mathbf{f}(\mathbf{x})) = \mathbf{0}$$

for all $\mathbf{x} \in X$.

This theorem tells us that $\mathbf{y} = \mathbf{f}(\mathbf{x})$ is implicitly defined by $\mathbf{\Phi}(\mathbf{x}, \mathbf{y}) = \mathbf{0}$. Then, if we differentiate $\mathbf{\Phi}$ with respect to $\mathbf{x}$ we get

$$\frac{\partial\mathbf{\Phi}}{\partial\mathbf{x}} + \frac{\partial\mathbf{\Phi}}{\partial\mathbf{f}}\frac{\mathbf{df}}{\mathbf{dx}} = 0.$$

From this expression we can notice that, by knowing $\mathbf{\Phi}$, we can compute the derivative of the function $\mathbf{f}$ with respect to $\mathbf{x}$, even though we do not have an explicit expression for it, that is,

$$\frac{\mathbf{df}}{\mathbf{dx}} = -\left(\frac{\partial \mathbf{\Phi}}{\partial \mathbf{y}}\right)^{-1} \frac{\partial \mathbf{\Phi}}{\partial \mathbf{x}}. \tag{2.22}$$

Next, $\mathbf{\Phi}$ can be computed as follows. If $\mathbf{y} = \mathbf{y}^*$ is a value where a cost function $C(\mathbf{x}, \mathbf{y})$ has a minimum, $\mathbf{\Phi}$ can be computed by the fact that, at the minimum of this cost function, $\frac{\partial C(\mathbf{x},\mathbf{y}^*)}{\partial \mathbf{y}} = 0$, then we choose $\mathbf{\Phi} = \frac{\partial C}{\partial \mathbf{y}}$. Thus, by the implicit function theorem, in a neighborhood of $\mathbf{y}^*$ the Jacobian of $\mathbf{f}$ is

$$\nabla \mathbf{f} = -\left(\frac{\partial^2 C}{\partial \mathbf{y}^2}\right)^{-1} \left(\frac{\partial^2 C}{\partial \mathbf{y} \partial \mathbf{x}}\right)^{\top} \tag{2.23}$$

This is the case when the function $\mathbf{f}$ is involved in a cost function with no constraints, otherwise, determining $\mathbf{\Phi}$ takes additional steps.

For the visual odometry process just described, the error propagation is performed in two steps. In the first, the covariance of each matching point is propagated through the least-squares stereo reconstruction process to get the covariance estimate of the corresponding 3D scene point. In the second step, the covariance of each 3D point of the two point clouds that are aligned are propagated through the pose estimation process to finally obtain the covariance of the relative pose measurement.

First order error propagation requires the derivatives of a function that converts matching points into 3D points in the first step, and 3D point clouds into a pose in the last step. Although we do not have access to an explicit function for each step, implicit functions are given by each of the involved minimization processes. Next we show how we compute the ensuing Jacobians.

2.3.1 Matching Point Error Propagation

We want the covariance $\mathbf{\Sigma}_p$ of the 3D scene point $\mathbf{p} = [x, y, z]^\top$ given the covariance $\mathbf{\Sigma}_m$ of the left image matching feature $\mathbf{m} = [u, v]^\top$ and the covariance $\mathbf{\Sigma}_{m'}$ of the right image matching feature $\mathbf{m}' = [u', v']^\top$. For instance, if we are using SIFT descriptors, the scale at which each feature was found can be used as an estimate for its covariance.

Next, to find $\mathbf{\Sigma}_p$ we need to obtain a first-order propagation of the covariance of the uncorrelated matching image feature, which is given by

$$\mathbf{\Sigma}_p = \nabla \mathbf{g} \begin{bmatrix} \mathbf{\Sigma}_m & \mathbf{0}_{2\times 2} \\ \mathbf{0}_{2\times 2} & \mathbf{\Sigma}_{m'} \end{bmatrix} \nabla \mathbf{g}^\top \tag{2.24}$$

where $\nabla \mathbf{g}$ is the Jacobian of the explicit function $\mathbf{g}$ that maps a pair of matching image points $\mathbf{u} = [u, v, u', v']^\top$ into its corresponding 3D scene point $\mathbf{p}$, i.e. $\mathbf{p} = \mathbf{g}(\mathbf{u})$.

As in this step the 3D scene point is found by solving the overdetermined system of equations given by Eq. 2.3, so as to apply the implicit function theorem, we need to express this process as an optimization problem. Thus, finding the 3D scene point $\mathbf{p}$ can be seen as minimizing the squared L_2-norm of the residual of Eq. 2.3, that is,

$$C(\mathbf{u}, \mathbf{p}) = \|\mathbf{A}\,\mathbf{p} - \mathbf{b}\|^2. \tag{2.25}$$

Computing the gradient of 2.25 with respect to $\mathbf{p}$ and setting it to zero, we find the minimum at

$$\mathbf{p}^* = (\mathbf{A}^\top \mathbf{A})^{-1} \mathbf{A}^\top \mathbf{b}, \tag{2.26}$$

assuming $\mathbf{A}$ to be invertible.

Lastly, having defined Eq. 2.25, by the implicit function theorem, the Jacobian of $\mathbf{g}$ is given by

$$\nabla \mathbf{g} = -\left(\frac{\partial^2 C}{\partial \mathbf{p}^2}\right)^{-1} \left(\frac{\partial^2 C}{\partial \mathbf{p} \partial \mathbf{m}}\right)^\top . \tag{2.27}$$

2.3.2 Point Cloud Error Propagation

In this step we are looking for the covariance $\boldsymbol{\Sigma}_d$ of the relative pose constraint $\mathbf{d}$ expressing the relative motion of the camera, given the covariances of each of the 3D points on the two point clouds. Here, again, this covariance will be computed by a first-order propagation, and we will need to compute the Jacobian of a function $\mathbf{h}$ that maps the points $\mathbf{P} = \left\{\mathbf{p}_1^{(i)}, \mathbf{p}_2^{(i)}\right\}$ on the two point clouds into the relative pose $\mathbf{d}$ that indicates the relative motion between the frame of the two clouds, i.e. $\mathbf{d} = \mathbf{h}(\mathbf{P})$.

If we express the relative pose $\mathbf{d}$ using Euler angles to represent its orientation, Eq. 2.4 can be written as follows

$$C(\mathbf{P}, \mathbf{d}) = \sum_{i=0}^{N} \left\| \mathbf{p}_1^{(i)} - \left(\mathbf{rot}(\phi_d, \theta_d, \psi_d)\, \mathbf{p}_2^{(i)} + \begin{bmatrix} x_d \\ y_d \\ z_d \end{bmatrix} \right) \right\|^2 , \tag{2.28}$$

where $\mathbf{d} = [x_d, y_d, z_d, \phi_d, \theta_d, \psi_d]^\top$, $\mathbf{rot}(\phi_d, \theta_d, \psi_d)$ is the rotation matrix defined by the Euler angles, and N is the point cloud size. The optimal value for $\mathbf{d}$ is computed with either one of the two approaches described in Sect. 2.2.

Thus, with the implicit function theorem the Jacobian of $\mathbf{d} = \mathbf{h}(\mathbf{P})$ is given by

$$\nabla \mathbf{h} = -\left(\frac{\partial^2 C}{\partial \mathbf{d}^2}\right)^{-1} \left(\frac{\partial^2 C}{\partial \mathbf{d} \partial \mathbf{P}}\right)^\top . \tag{2.29}$$

Finally, the covariance $\boldsymbol{\Sigma}_d$ of the relative pose constraint $\mathbf{d}$ will be given by,

$$\boldsymbol{\Sigma}_d = \nabla \mathbf{h}\, \boldsymbol{\Sigma}_P\, \nabla \mathbf{h}^\top, \tag{2.30}$$

where $\boldsymbol{\Sigma}_P$ is the covariance of the two clouds of points $\mathbf{P}$, that is,

$$\boldsymbol{\Sigma}_P = \text{diag}\left(\boldsymbol{\Sigma}_{p1}^{(1)}, \ldots, \boldsymbol{\Sigma}_{p1}^{(N)}, \boldsymbol{\Sigma}_{p2}^{(1)}, \ldots, \boldsymbol{\Sigma}_{p2}^{(N)}\right), \tag{2.31}$$

which is a block diagonal matrix, where $\boldsymbol{\Sigma}_{p1}^{(i)}$ and $\boldsymbol{\Sigma}_{p2}^{(i)}$ are the covariances of the i-th point of the first and second clouds, respectively.

An alternative procedure would be to rely on optimization approaches to obtain the uncertainty in the pose estimation, similarly to [15]. However, in this work we opted instead for the use of the implicit function theorem to propagate uncertainties as it yields closed-form expressions.

2.3.3 Error Propagation Tests

The following tests evaluate whether the covariance resulted from the error propagation is consistent with N Monte Carlo runs, using both synthetic and real data. To this end, we compute the normalized state estimation error squared or NEES [16] for each Monte Carlo run

$$\epsilon_i = [\mathbf{s}_i - \mu]^\top\ \boldsymbol{\Sigma}^{-1}\ [\mathbf{s}_i - \mu] \tag{2.32}$$

and take the average

$$\bar{\epsilon} = \frac{1}{N}\sum_{i=0}^{N} \epsilon_i, \tag{2.33}$$

where $\mathbf{s}_i$ is the result of a Monte Carlo run, $\boldsymbol{\Sigma}$ the covariance obtained with the error propagation and μ is the solution to either Eq. 2.3, for the matching point error propagation, or Eq. 2.4, for the point cloud error propagation.

If the Monte Carlo runs are consistent with the error propagation results, then $N\bar{\epsilon}$ will have a Chi-Squared density with Nn_x degrees of freedom or $\chi^2_{Nn_x}$, where n_x is the dimension of $\mathbf{s}_i$ and χ^2_n denotes a Chi-Squared distribution of n degrees of freedom. We validate this using a Chi-square test with a two sided 95% probability region, defined by the interval $[l_1, l_2]$. Thus, if

$$N\bar{\epsilon} \in [l_1, l_2] \tag{2.34}$$

we confirm that the error propagation result is consistent with the Monte Carlo runs.

2.3.3.1 Synthetic Data

To test the matching point error propagation, we simulated a stereo camera system using the left camera as the reference frame, where the relative pose of the right camera with respect to the reference frame is given by a translation vector of $[0.385\,\text{m}, 0.007\,\text{m}, 0.00\,\text{m}]^\top$ and a rotation matrix expressed in Euler angles of $\mathbf{rot}(0.040\,\text{rad}, 0.016\,\text{rad}, 0.014\,\text{rad})$. Then, we simulated a ground truth 3D scene point and its corresponding imaged points in both cameras.

Next, we set the covariance $\boldsymbol{\Sigma}_m = \boldsymbol{\Sigma}_{m'} = \text{diag}(2\,\text{px}, 2\,\text{px})^2$ for both imaged points and apply the first-order error propagation (Eq. 2.24) to such covariances. Then, we perform a Monte Carlo simulation by generating a set of 300 pairs of random matching points around such image points and for each sample we obtain its corresponding 3D point with Eq. 2.26.

Figure 2.2 shows the simulated samples, the Monte Carlo covariance (black line), and the covariance computed with the error propagation (dashed green line). All hyperellipsoids represent iso-uncertainty curves plotted at a scale of 2 standard deviations.

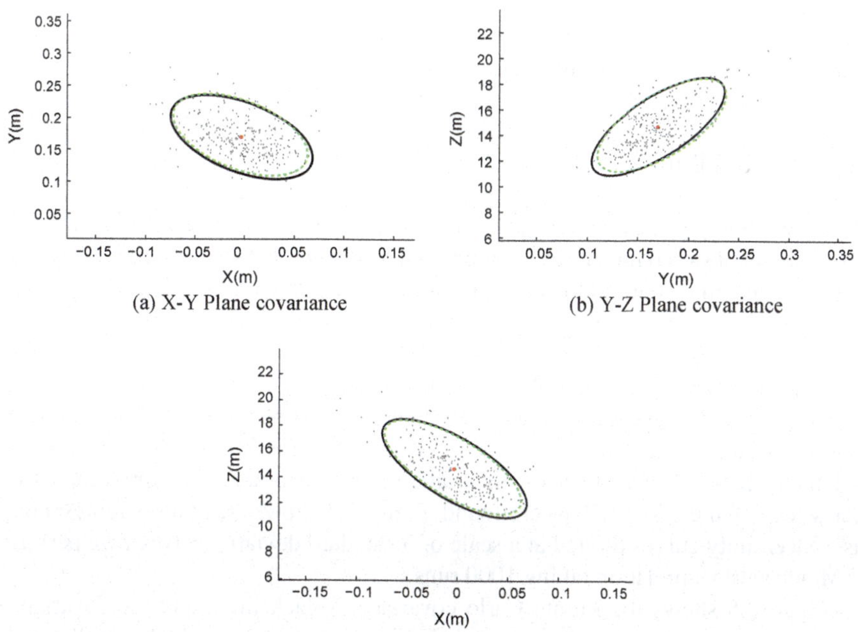

(a) X-Y Plane covariance

(b) Y-Z Plane covariance

(c) X-Z Plane covariance

Fig. 2.2 Simulation of error propagation for the stereo reconstruction of a single image pair. The covariance obtained by Monte Carlo simulation is represented by the *black ellipse*, while the covariance computed with the first-order error propagation is plotted with the dashed *green ellipse*. All hyperellipsoids represent iso-uncertainty curves plotted at a scale of 2 standard deviations. The red point shows the mean of reconstructed 3D point

This test yielded $N\bar{\epsilon} = 860.1563$, lying within the interval [831.3, 970.4], which defines the two-sided 95% probability region for a χ^2_{900} variable, thus confirming the consistency of the error propagation.

To test the error propagation of the pose estimation process, we simulated two stereo systems with a known relative pose and placed 100 scene points uniformly distributed in the field of view of the four cameras and compute their corresponding imaged points, assigning to such points a covariance of $\boldsymbol{\Sigma}_m = \boldsymbol{\Sigma}_{m'} = \text{diag}(2\,\text{px}, 2\,\text{px})^2$. Then, we propagate their covariance along the whole visual odometry process.

Next, to perform a Monte Carlo simulation, with the covariance of each imaged points, we generate 1000 samples around each point, which yields 1000 point clouds. Then, we apply the least-squares best-fit pose transformation to each cloud.

Figure 2.3 shows the Monte Carlo covariance (black line), and the covariance computed with the implicit theorem function (dashed green line). All hyperellipsoids represent iso-uncertainty curves plotted at a scale of 2 standard deviations. From both Figs. 2.2 and 2.3, we can note that the covariances obtained with the first order error propagation are similar to covariances computed with Monte Carlo, with the advantage of the former being less expensive.

For this test we get $N\bar{\epsilon} = 6.07 \times 10^3$, which confirmed the consistency of the error propagation since it lies within the interval $\left[5.8 \times 10^3, 6.2 \times 10^3\right]$, defining the two-sided 95% probability region for a χ^2_{6000} variable.

2.3.3.2 Real Data

Next, we show a test using real data. To this end, we took a pair of stereo images out of a data set of stereo images in an outdoor environment. In the stereo camera system we used the left camera as the reference frame, as a result of the calibration process the relative pose of the right camera with respect to the left camera is given by a translation vector of $[0.385\,\text{m}, 0.007\,\text{m}, 0.00\,\text{m}]^\top$ and a rotation matrix expressed in Euler angles of $\mathbf{rot}(0.040\,\text{rad}, 0.016\,\text{rad}, 0.014\,\text{rad})$.

From these images we compute the relative motion between the stereo cameras and propagate the uncertainty of each matching feature through the whole visual odometry process. For our tests, the value of this covariance is approximated by the scale at which the SIFT where found. Figure 2.4 shows the ellipses representing iso-uncertainty curves plotted at a scale of 3 standard deviations. Next, we perform a Monte Carlo simulation taking 1000 runs.

Figure 2.5 shows the Monte Carlo covariance (black line), and the covariance computed with the implicit theorem function (dashed green line). All hyperellipsoids represent iso-uncertainty curves plotted at a scale of 2 standard deviations.

For this test we get $N\bar{\epsilon} = 6.14 \times 10^3$, lying in the aforementioned interval that defined the two-sided 95% probability region for a χ^2_{6000} variable. Thus confirming the consistency of the error propagation results with the Monte Carlo runs.

The above experiments showed the consistency of using a closed form expression to compute the covariance of the relative motion measurements obtained with stereo

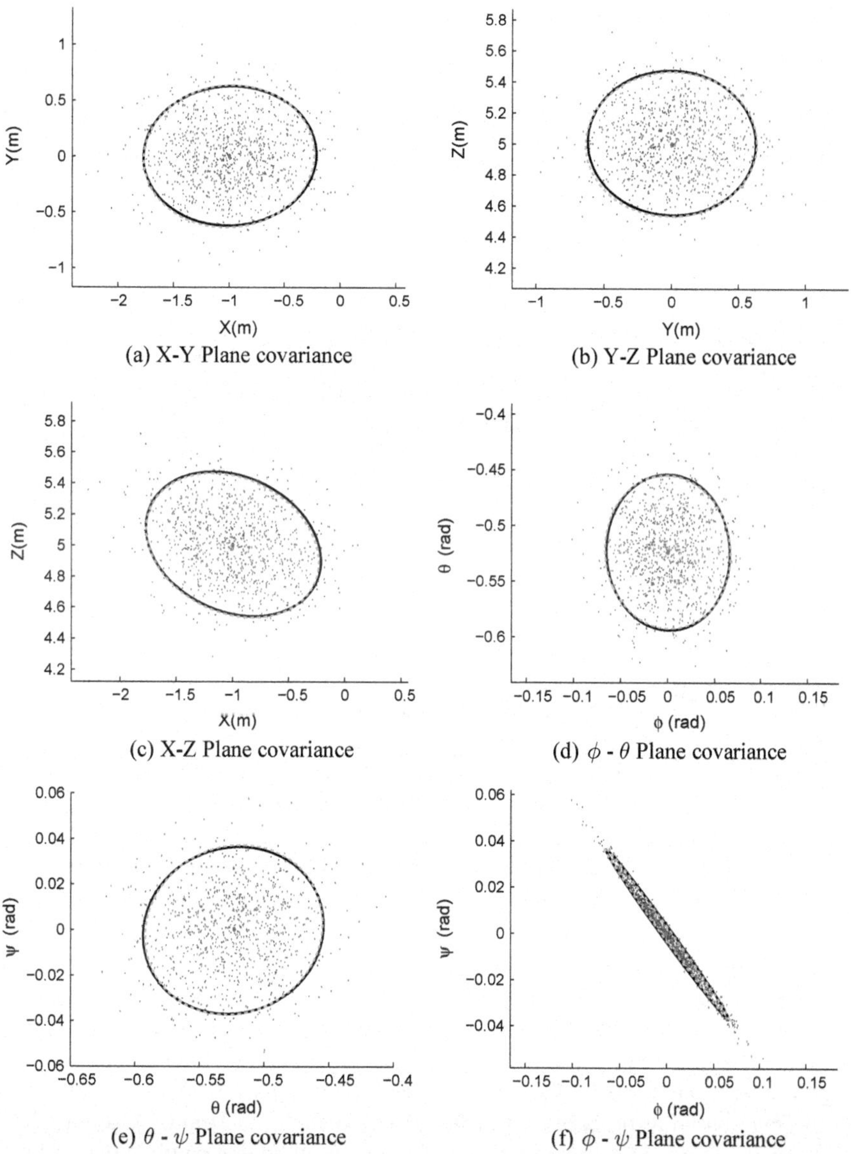

(a) X-Y Plane covariance

(b) Y-Z Plane covariance

(c) X-Z Plane covariance

(d) ϕ - θ Plane covariance

(e) θ - ψ Plane covariance

(f) ϕ - ψ Plane covariance

Fig. 2.3 Simulation of the error propagation of the pose estimation from the two point clouds. The covariance obtained by Monte Carlo simulation is respresent by the *black ellipse*, while the covariance computed with the implicit function theorem is plotted with the dashed *green ellipse*. All hyperellipsoids represent iso-uncertainty curves plotted at a scale of 2 standard deviations. The *red point* shows the mean of the estimated pose

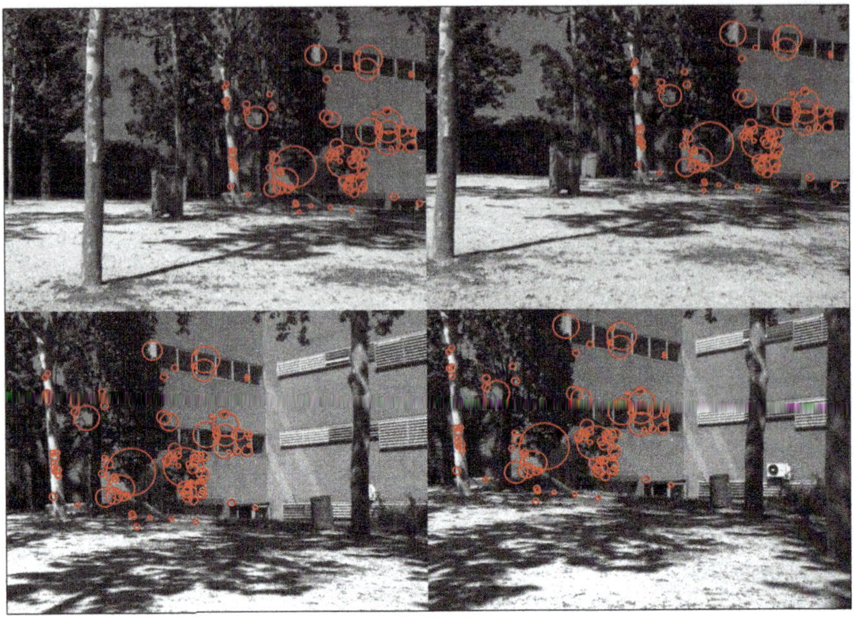

Fig. 2.4 Some of the SIFT correspondences in two consecutive stereo image pairs and their covariance. The ellipses represent iso-uncertainty curves plotted at a scale of 3 standard deviations

cameras. An alternative technique to model the error in motion measurements can be to use the method introduced by Matthies and Shafer in [15]. In their approach when the motion is purely translational, their technique exploits the linearity of such a case and obtains a direct solution, however, when the motion involves rotation the approach has to resort to an iterative solution and applies the Gauss-Newton method to refine iteratively the estimate.

Nevertheless, in some robotic applications it is desirable to use instead a closed-form expression to compute the covariance of the relative motion measurements, regardless of the type of motion. For instance, in map building applications modeling the uncertainty with a closed form expression is preferable. Although back-end SLAM approaches have evolved to the point of being very efficient, the bottle-neck of the map building process is still the front-end SLAM. Data association strategies usually require to compute relative motion measurements multiple times, thus a closed form approach would be preferable over iterative methods. For instance, for the case of laser range finders, a widely used Iterative Closest Point implementation in robotic applications introduced in [17] computes the covariance of the relative motion estimates with a similar closed form formulation to the one described in this Chapter.

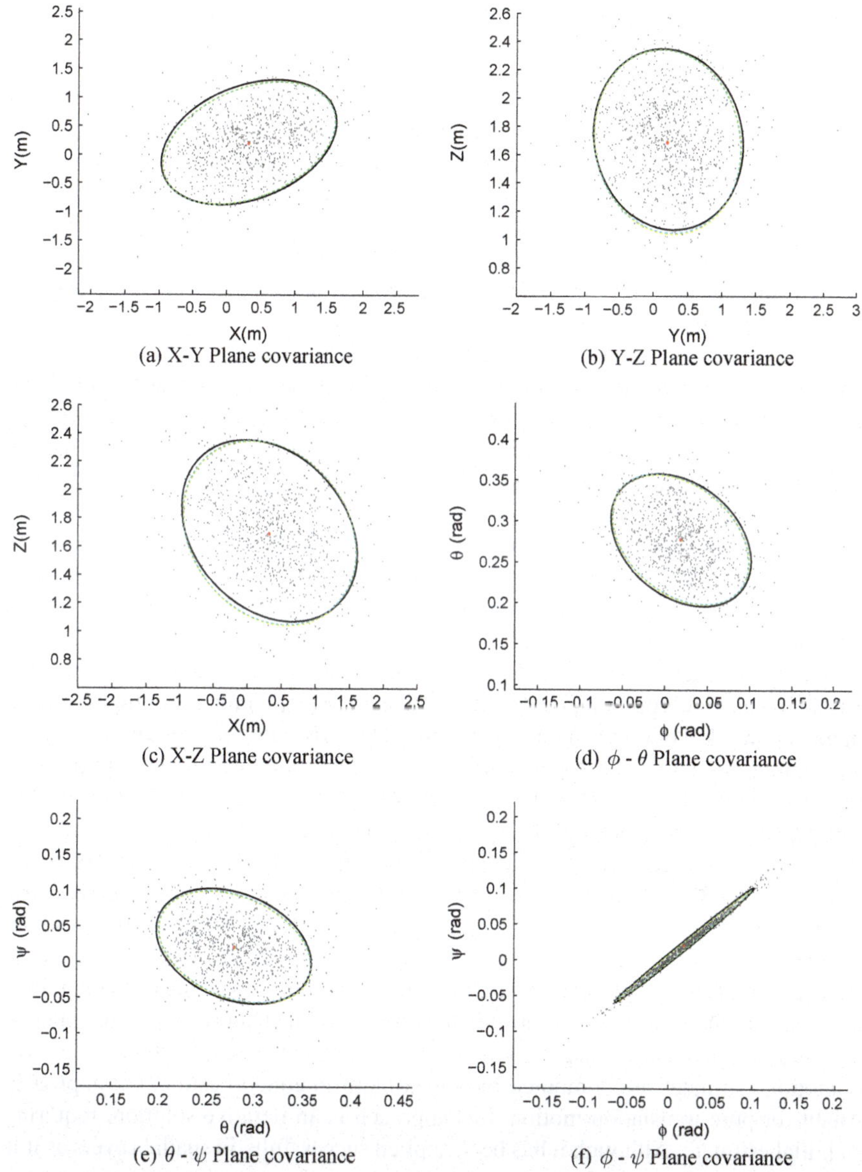

(a) X-Y Plane covariance

(b) Y-Z Plane covariance

(c) X-Z Plane covariance

(d) ϕ - θ Plane covariance

(e) θ - ψ Plane covariance

(f) ϕ - ψ Plane covariance

Fig. 2.5 Error propagation of the relative pose estimation between two robot poses using stereo images. The covariance obtained by Monte Carlo simulation is respresent by the *black ellipse*, while the covariance computed with the implicit function theorem is plotted with the dashed *green ellipse*. All hyperellipsoids represent iso-uncertainty curves plotted at a scale of 2 standard deviations. The red point shows the mean of the estimated pose

2.4 Bibliographical Notes

Visual odometry is the problem of estimating the egomotion of the robot from one or multiple cameras attached to it. The term visual odometry was introduced in [18] but the problem of estimating the motion of a vehicle from visual imagery was previously addressed by Moravec in [19]. Visual odometry is a particular case of Structure From Motion or SFM [20], SFM seeks to recover the camera pose and three-dimensional structure from a set of images. It takes the 2D information and recover the original 3D information, inverting the effect of the projection process. Bundle adjustment (BA) is frequently used to improve upon SFM solutions [12].

Unlike SFM in visual odometry the 3D motion is estimated as a new frame arrives. The works on visual odometry have evolved in two branches, that is, in monocular and stereo visual odometry. Besides this, each work distinguishes in the way they solve each part of the problem, i.e., feature detection, feature matching, motion estimation, and local optimization. A comprehensive review on the evolution of these works appeared in [2, 3].

The work presented in this Chapter belongs to the stereo-based approaches to visual odometry. Most of these approaches track features along the frames obtained by corner detectors [19], such as Forstner [21, 22], Harris [23, 24] or Shi-Tomasi [25] or by selecting key points after performing dense stereo [26]. In these works 3D points are triangulated for every stereo pair, which are used to solve the motion estimation and incorporating RANSAC in the motion estimation for outlier removal. In the landmark work presented in [18], instead of tracking features among images the features are detected independently in all images. Additionally, instead of using corner detector, blob detectors has been also employed, such as SIFT in [27, 28], CENSURE in [29], SURF in [30], or BRIEF [31] descriptors in [32].

In visual odometry, errors in locating the image features lead to errors in the location of the 3D feature points after stereo reconstruction, which eventually cause errors in motion estimates. One of the first error modeling approaches for visual odometry was presented by Matthies and Shafer in [15]. In this work, the errors in the location of the image features are modeled as random variables following a 3D Gaussian distribution. Thus, the error in the location of 3D feature points are obtained by a first-order linear propagation of the covariances of image features. The error in motion estimates are computed by a maximum-likelihood estimation approach. Albeit for pure translation motion, this approach is an iterative solution, requiring an initial estimate. Although it has been applied successfully in spatial rovers as it is reported in [24], a closed-form expression for the motion uncertainty is preferred.

The implicit function theorem was initially exploited in [33, 34], where Chowdhury and Chellappa derive analytical expressions for the covariance of structure and motion estimates as a function of the covariance of the image correspondences. In [35], in order to estimate the uncertainty of the robot pose obtained via a correspondence-based method with stereo images, the authors employ the implicit function theorem to derive the pose uncertainty from a maximum likelihood formulation.

Besides visual odometry, the modeling of errors in pose estimation has been also addressed for wheel odometry [36–38] as well as for other sensor modalities. For range sensors, a model of the error of the pose estimation obtained with ICP is presented in [17] for laser range finders and in [39] for time-of-flight sensors. Both approaches are strongly related to the approach presented in this Chapter as they also model motion uncertainty by a first-order error propagation based on the implicit function theorem.

References

1. P.J. Besl, N.D. McKay, A method for registration of 3D shapes. IEEE Trans. Pattern Anal. Mach. Intell. **14**(2), 239–256 (1992). Feb
2. D. Scaramuzza, F. Fraundorfer, Visual odometry: Part I—the first 30 years and fundamentals. Robot. Automat. Mag. **18**(4), 80–92 (2011). Dec
3. F. Fraundorfer, D. Scaramuzza, Visual odometry: Part II—matching, robustness, and applications. Robot. Automat. Mag. **19**(2), 78–90 (2012)
4. V. Ila, J. Andrade-Cetto, R. Valencia, A. Sanfeliu, Vision-based loop closing for delayed state robot mapping, in *Proceedings of IEEE/RSJ International Conference on Intelligent Robots Systems* (San Diego, 2007), pp. 3892–3897
5. D.G. Lowe, Distinctive image features from scale-invariant keypoints. Int. J. Comput. Vis. **60**(2), 91–110 (2004). Nov
6. M. Fischler, R. Bolles, Random sample consensus: A paradigm for model fitting with applications to image analysis and automated cartography. Comm. ACM **24**, 381–385 (1981)
7. O. Faugeras, *Three-Dimensional Computer Vision: A Geometric Viewpoint* (The MIT Press, Cambridge, 1993)
8. C.G. Harris, M. Stephens, A combined corner edge detector, in *Proceedings of Alvey Vision Conference* (Manchester, 1988), pp. 189–192
9. J. Shi, C. Tomasi, Good features to track, in *Proceedings of 9th IEEE Conference on Computer Vision and Pattern Recognition* (Seattle, 1994), pp. 593–600
10. A.J. Davison, I.D. Reid, N.D. Molton, O. Stasse, MonoSLAM: Real-time single camera SLAM. IEEE Trans. Pattern Anal. Mach. Intell. **26**(6), 1052–1067 (2007)
11. K. Mikolajczyk, C. Schmid, A performance evaluation of local descriptors. IEEE Trans. Pattern Anal. Mach. Intell. **27**(10), 1615–1630 (2005)
12. R. Hartley, A. Zisserman, *Multiple View Geometry in Computer Vision* (Cambridge University Press, Cambridge, 2000)
13. B.K.P. Horn, H.M. Hilden, S. Negahdaripour, Closed-form solution of absolute orientation using orthonormal matrices. J. Opt. Soc. Am. A **5**(7), 1127–1135 (1988)
14. K.S. Arun, T.S. Huang, S.D. Blostein, Least-squares fitting of two 3-D point sets. IEEE Trans. Pattern Anal. Mach. Intell. **9**(5), 698–700 (1987). Sep
15. L. Matthies, S. Shafer, Error modeling in stereo navigation. IEEE J. Robot. Autom. **3**(3), 239–248 (1987). Jun
16. Y. Bar-Shalom, X. Rong Li, T. Kirubarajan, *Estimation with Applications to Tracking and Navigation* (Wiley, New York, 2001)
17. A. Censi, An accurate closed-form estimate of ICP's covariance, in *Proceedings of the IEEE International Conference on Robotics and Automation* (Rome, 2007), pp. 3167–3172
18. D. Nister, O. Naroditsky, J. Bergen, Visual odometry, in *Proceedings of 18th IEEE Conference on Computer Vision and Pattern Recognition* (Washington, 2004), pp. 652–659
19. H.P. Moravec, Obstacle avoidance and navigation in the real world by a seeing robot rover. Ph.D. thesis, Stanford University, California, 1980

20. H.C. Longuet-Higgins, A computer program for reconstructing a scene from two projections. Nature **293**(11), 133–135 (1981)
21. C. Olson, L. Matthies, M. Schoppers, M. Maimone, Robust stereo ego-motion for long distance navigation, in *Proceedings of 14th IEEE Conference on Computer Vision and Pattern Recognition*, vol. 2 (Hilton Head, SC, 2000), pp. 453–458
22. C. Olson, L. Matthies, M. Schoppers, M. Maimone, Rover navigation using stereo ego-motion. Robot. Auton. Syst. **43**(4), 215–229 (2003)
23. Y. Cheng, M. Maimone, L. Matthies, Visual odometry on the mars exploration rovers—a tool to ensure accurate driving and science imaging. Robot. Automat. Mag. **13**(2), 54–62 (2006)
24. M. Maimone, Y. Cheng, L. Matthies, Two years of visual odometry on the mars exploration rovers. J. Field Robot. **24**(3), 169–186 (2007)
25. A. Milella, R. Siegwart, Real-time scattering compensation for time-of-flight camera, in *International Conference on Computer Vision Systems* (2006), pp. 21–24
26. S. Lacroix, A. Mallet, R. Chatila, L. Gallo, Real-time scattering compensation for time-of-flight camera, in *Proceedings of International Symposium on Artificial Intelligence, Robotics and Automation in Space* (Noordwijk, 1999), pp. 433–440
27. S. Se, D. Lowe, J. Little, Mobile robot localization and mapping with uncertainty using scale-invariant visual landmarks. Int. J. Robot. Res. **21**(8), 735–758 (2002). Aug
28. S. Se, D. Lowe, J. Little, Vision-based global localization and mapping for mobile robots. IEEE Trans. Rob. **21**(3), 364–375 (2005). Jun
29. K. Konolige, M. Agrawal, J. Solà, Large scale visual odometry for rough terrain, in *Proceedings of 13th International Symposium of Robotics Research* (Hiroshima, 2007)
30. L. Kneip, M. Chli, R. Siegwart, Robust real-time visual odometry with a single camera and an IMU, in *Proceedings of British Machine Vision Conference* (BMVA Press, Dundee, 2011), pp. 20.1–20.10
31. M. Calonder, V. Lepetit, C. Strecha, P. Fua, BRIEF: binary robust independent elementary features, in *Proceedings of 11th European Conference on Computer Vision* (2010), pp. 778–792
32. T. Kazik, L. Kneip, J. Nikolic, M. Pollefeys, R. Siegwart, Real-time 6D stereo visual odometry with non-overlapping fields of view, in *IEEE Conference on Computer Vision and Pattern Recognition* (Providence, 2012), pp. 1529–1536
33. A.R. Chowdhury, R. Chellappa, Statistical error propagation in 3D modeling from monocular video, in *IEEE CVPR Workshops* (Madison, 2003)
34. A.R. Chowdhury, R. Chellappa, Stochastic approximation and rate-distortion analysis for robust structure and motion estimation. Int. J. Comput. Vis. **55**(1), 27–53 (2003)
35. P. Zhang, J. Gu, E. Milios, Registration uncertainty for robot self-localization in 3D, in *Proceedings of 2nd Canadian Conference on Computer and Robot Vision* (Victoria, 2005), pp. 490–497
36. J. Borenstein, L. Feng, Measurement and correction of systematic odometry errors in mobile robots. IEEE Trans. Rob. **12**(6), 869–880 (1996). Dec
37. A. Martinelli, The odometry error of a mobile robot with a synchronous drive system. IEEE Trans. Rob. **18**(3), 399–405 (2002). Jun
38. J.M. Mirats, J.L. Gordillo, C.A. Borja, A closed-form expression for the uncertainty in odometry position estimate of an autonomous vehicle. IEEE Trans. Rob. **21**(5), 1017–1022 (2005). Oct
39. S. Foix, G. Alenyà, J. Andrade-Cetto, C. Torras, Object modeling using a ToF camera under an uncertainty reduction approach, in *Proceedings of IEEE International Conference on Robotics and Automation* (Anchorage, 2010), pp. 1306–1312

Chapter 3
SLAM Back-End

The SLAM problem has been traditionally addressed as a state estimation problem in which perception and motion uncertainties are coupled. The standard probabilistic solution to SLAM maintains an estimate of the most recent robot pose and the locations of environmental landmarks with an Extended Kalman filter (EKF), a solution usually referred to as EKF-SLAM [1, 2]. This solution has allowed to identify the basic properties of such a coupled state estimation problem [3, 4]; however, it presents drawbacks with respect to precision due to linearizations and scalability.

One attempt to improve scalability is the use of information-based representations [5–8]. When estimating both the last robot pose and features, the resulting information matrix turns out to be approximately sparse with small matrix entries for distant landmarks [5]. Furthermore, exactly sparse information matrices can be obtained by estimating the entire robot path along with the map, a solution usually referred to as full SLAM [8–10]. Going a step further, exact sparsity can also be achieved by estimating only the robot trajectory [6, 11, 12]. However, estimating only the history of robot poses also presents some drawbacks. By estimating all the robot poses, the state grows independently of the size of environment and adding all possible links reduces the sparsity of the information matrix. Moreover, when using a linearized approach, the accumulation of linearization errors introduced by each new link produces overconfident estimates, which leads to filter inconsistency [13, 14]. Fortunately, a principled solution to these issues is given by the Pose SLAM approach [11].

Pose SLAM is the variant of SLAM where only the robot trajectory is estimated with an information-based representation. However, it only keeps non-redundant poses and highly informative links, thus building a more compact map that translates into a significant reduction of the computational cost and a delay of the filter inconsistency, maintaining the quality of the estimation for longer mapping sequences.

In this work we use Pose SLAM as the basic state estimation machinery. The original Pose SLAM algorithm [11] was developed for SE(2) and in this Chapter we

R. Valencia and J. Andrade-Cetto, *Mapping, Planning and Exploration with Pose SLAM*, Springer Tracts in Advanced Robotics 119,
DOI 10.1007/978-3-319-60603-3_3

complement this formulation with a Pose SLAM implementation in SE(3) using both Euler and quaternion parameterizations for rotations. We begin this Chapter with an explanation in Sect. 3.1 of the main parts that comprise the Pose SLAM approach and explain the data association and map management techniques introduced in [11], along with our proposed loop closure strategy [7]. Next, our 6 DOF Pose SLAM implementation is presented in Sect. 3.2 and in Sect. 3.3 we introduce our method to compute traversability maps. Up to this point in the Chapter our exposition is sensor agnostic, i.e. we do not assume that any specific sensor modality is used. We change this in Sect. 3.4, when we show specific mapping results for the 2D Pose SLAM case using stereo images and for the Pose SLAM extension in 6 DOF, using a 3D-Laser range finder to build a volumetric map and a traversability map tailored to be used by a heterogeneous fleet of service robots in urban settings [15]. Lastly, in Sect. 3.5 we give bibliographical notes.

3.1 Pose SLAM Preliminaries

The purpose of Pose SLAM is to compute an estimate of the robot trajectory $\mathbf{x}_k = [x_0^\top \; \dots \; x_k^\top]^\top$, with each x_i a random vector corresponding to the i-th robot pose, given the history of proprioceptive observations $\mathcal{Z}_k$ and motion commands $\mathcal{U}_k$. The robot trajectory $\mathbf{x}_k$ is maintained with the canonical representation of the Gaussian distribution, that is,

$$\begin{aligned} p(\mathbf{x}_k | \mathcal{Z}_k, \mathcal{U}_k) &= \mathcal{N}(\mathbf{x}_k; \boldsymbol{\mu}_k, \boldsymbol{\Sigma}_k) \\ &= \mathcal{N}^{-1}(\mathbf{x}_k; \boldsymbol{\eta}_k, \boldsymbol{\Lambda}_k), \end{aligned} \tag{3.1}$$

with

$$\boldsymbol{\Lambda}_k = \boldsymbol{\Sigma}_k^{-1} \text{ and } \boldsymbol{\eta}_k = \boldsymbol{\Sigma}_k^{-1} \boldsymbol{\mu}_k, \tag{3.2}$$

where $\boldsymbol{\mu}_k$ is the mean state vector and $\boldsymbol{\Sigma}_k$ its covariance matrix. $\boldsymbol{\Lambda}_k$ and $\boldsymbol{\eta}_k$ are the information matrix and information vector, respectively.

The on-line form of Pose SLAM [6, 11] computes this estimate incrementally by performing a state augmentation and a state update operation at each iteration. Next, we describe these operations as well as the data association and map management processes.

3.1.1 State Augmentation

This operation augments the state vector to contain a new pose. That is, given the trajectory state $\mathbf{x}_{k-1}$ and all measurements $\mathcal{Z}_{k-1}$ and control inputs $\mathcal{U}_{k-1}$ up to time $k-1$, the execution of action u_k augments the trajectory state with a new pose x_k, obtaining the distribution $p(\mathbf{x}_{k-1}, x_k|\mathcal{Z}_{k-1}, \mathcal{U}_k)$ for the trajectory state $\mathbf{x}_k = [\mathbf{x}_{k-1}^\top, \; x_k^\top]^\top$, with $\mathcal{U}_k$ the set of control inputs up to time k. By the Markov assumption, the posterior can be factored as [6]

$$p(\mathbf{x}_{k-1}, x_k|\mathcal{Z}_{k-1}, \mathcal{U}_k) = p(x_k|\mathbf{x}_{k-1}, u_k) \; p(\mathbf{x}_{k-1}|\mathcal{Z}_{k-1}, \mathcal{U}_{k-1}), \tag{3.3}$$

with $p(x_k|\mathbf{x}_{k-1}, u_k)$ the probabilistic state transition model and $p(\mathbf{x}_{k-1}|\mathcal{Z}_{k-1}, \mathcal{U}_{k-1})$ our prior distribution.

In Pose SLAM, the state transitions model results from the composition of a motion command u_k to the previous pose x_{k-1},

$$x_k = f(x_{k-1}, u_k) + \mathbf{w}_k \tag{3.4}$$

$$\approx f(\mu_{k-1}, \mu_u) + \mathbf{F}_k \, (\mathbf{x}_{k-1} - \boldsymbol{\mu}_{k-1}) + \mathbf{w}_k, \tag{3.5}$$

where $f(x_{k-1}, u_k) = x_{k-1} \oplus u_k$, with $\oplus$ the operator used to add the relative displacement u_k to pose x_{k-1}, as described in [16]. $\mathbf{F}_k$ is the Jacobian of f with respect to x_{k-1} evaluated at the mean μ_{k-1} and $\mathbf{w}_k = \mathcal{N}(0, \boldsymbol{\Sigma}_u)$ the white process noise.

With the linearized state transition model (3.5), the augmented state distribution (3.3) in information form is given by

$$\bar{\boldsymbol{\eta}}_k = \begin{bmatrix} \boldsymbol{\eta}_{1:k-2} \\ \eta_{k-1} - \mathbf{F}_k{}^\top \boldsymbol{\Sigma}_u^{-1} \, (f(\mu_{k-1}, u_k) - \mathbf{F}_k \, \mu_{k-1}) \\ \boldsymbol{\Sigma}_u^{-1} \, (f(\mu_{k-1}, u_k) - \mathbf{F}_k \, \mu_{k-1}) \end{bmatrix} \tag{3.6}$$

and

$$\bar{\boldsymbol{\Lambda}}_k = \begin{bmatrix} \boldsymbol{\Lambda}_{1:k-2,1:k-2} & \boldsymbol{\Lambda}_{1:k-2,k-1} & \mathbf{0} \\ \boldsymbol{\Lambda}_{k-1,1:k-2} & \Lambda_{k-1\,k-1} + \mathbf{F}_k^\top \boldsymbol{\Sigma}_u^{-1} \mathbf{F}_k & -\mathbf{F}_k^\top \boldsymbol{\Sigma}_u^{-1} \\ \mathbf{0} & -\boldsymbol{\Sigma}_u^{-1} \mathbf{F}_k & \boldsymbol{\Sigma}_u^{-1} \end{bmatrix} \tag{3.7}$$

where η_{k-1} and $\boldsymbol{\Lambda}_{k-1\,k-1}$ are used to denote the blocks of $\boldsymbol{\eta}_{k-1}$ and $\boldsymbol{\Lambda}_{k-1}$ corresponding to the $(k-1)$-th pose, and $\boldsymbol{\eta}_{1:k-2}$ and $\boldsymbol{\Lambda}_{1:k-2,1:k-2}$ indicate those ranging from the first to the $(k-2)$-th pose [11].

Augmenting the state in information form introduces shared information only between the new robot pose x_k and the previous one x_{k-1}, resulting in an information matrix with a tridiagonal block structure. If the state mean is available, this operation can be performed in constant time.

3.1.2 State Update

Sensor measurements in Pose SLAM are observations about the relative change between the current robot pose with respect to any of the previous poses kept in the trajectory estimate. This introduces shared information, but now between non-consecutive poses. The measurement model for the relative pose constraints is

$$z_{ki} = h(x_k, x_i) + \mathbf{v}_k \tag{3.8}$$

$$\approx h(\mu_k, \mu_i) + \mathbf{H}(\mathbf{x}_k - \boldsymbol{\mu}_k) + \mathbf{v}_k, \tag{3.9}$$

where $h(x_k, x_i) = \ominus x_k \oplus x_i$, that is, the tail-to-tail operation defined in [16], which computes the relative motion from x_k to x_i in the frame of reference of x_k. With white measurement noise $\mathbf{v}_k = \mathcal{N}(0, \boldsymbol{\Sigma}_z)$. Equation (3.9) is the first order linearized form of Eq. (3.8), with

$$\mathbf{H} = [\mathbf{0} \ldots \mathbf{0}\, \mathbf{H}_i\, \mathbf{0} \ldots \mathbf{0}\, \mathbf{H}_k], \tag{3.10}$$

the Jacobian of $h(x_k, x_i)$, where $\mathbf{H}_i$ and $\mathbf{H}_k$ are the Jacobians of h with respect to x_i and x_k, respectively.

Having defined the measurement model, the state is updated with the new measurement z_{ki} by the Extended Information Filter (EIF) update expressions [5], which add the following increments to $\bar{\eta}_k$ and $\bar{\boldsymbol{\Lambda}}_k$, respectively,

$$\Delta\eta = \mathbf{H}^\top \boldsymbol{\Sigma}_z^{-1}(z_{ki} - h(\mu_k, \mu_i) + \mathbf{H}\,\boldsymbol{\mu}_k) \tag{3.11}$$

$$\Delta\boldsymbol{\Lambda} = \mathbf{H}^\top \boldsymbol{\Sigma}_z^{-1} \mathbf{H}, \tag{3.12}$$

that is,

$$\eta_k = \bar{\eta}_k + \Delta\eta$$

and

$$\boldsymbol{\Lambda}_k = \bar{\boldsymbol{\Lambda}}_k + \Delta\boldsymbol{\Lambda}. \tag{3.13}$$

When establishing such a link, the update operation only modifies the diagonal blocks i and k of the information matrix $\boldsymbol{\Lambda}$ and introduces new off-diagonal blocks at locations ik, and ki. This operation is also executed in constant time, assuming the state mean to be available. These links enforce graph connectivity, or loop closure in SLAM parlance, and revise the entire path state estimate, reducing overall uncertainty.

3.1.3 Data Association

In the context of Pose SLAM, data association refers to the process of loop closure detection. Two phases can be distinguished during this process. First, we must to detect the possibility of a loop closure event. Second, we need to certify the presence of such a loop closure from sensor data. It is convenient to hypothesize whether a candidate link is informative enough before actually aligning the sensor readings since sensor registration is an expensive process.

Next, we show two different data association strategies for Pose SLAM that exploit the filter information to constrain the search for sensory matches only to a small number of neighboring poses. Both strategies work independently and represent two different ways of solving the data association problem in Pose SLAM, but can be directly applied to other pose-based SLAM techniques. For consistent estimates, both strategies are more efficient and less affected by perceptual aliasing, as opposed to data association techniques independent of the filter estimates, which directly search for feature matches in a sensor database [17].

In the first strategy [7], the set of candidate loop closing poses is chosen assuming independence amongst them by measuring the closeness of distributions from their Mahalanobis distance and selecting the most informative ones by computing their Bhattacharyya distance. In the second [11], the assumption of independence between poses is no longer needed and the set of candidate poses consists of poses with high probability of being close to each other, while the final candidates are selected by their information content, measured with the mutual information gain, thus considering the effect of the candidate link information gain on the whole state. These data association methods assume we have access to consistent estimates as they are less affected by perceptual aliasing, however, in case the estimates become inconsistent the data association process can be complemented with an appearance-based approach [18, 19].

3.1.3.1 Independent Measure of Information Content Between Poses

A comparison of the current pose estimate with the history of poses can tell whether the robot is in the vicinity of a previously visited place. This vicinity is measured computing the Mahalanobis distance from the prior estimate to all previously visited locations, i.e., for all $0 < i < k$,

$$d_M^2 = (\mu_k - \mu_i)^\top \left(\frac{\mathbf{\Sigma}_{kk} + \mathbf{\Sigma}_{ii}}{2} \right)^{-1} (\mu_k - \mu_i), \tag{3.14}$$

where $\mathbf{\Sigma}_{kk}$ and $\mathbf{\Sigma}_{ii}$ are the marginal covariances of poses k and i, with state means μ_k and μ_i, respectively.

An exact computation of $\mathbf{\Sigma}_{kk}$ and $\mathbf{\Sigma}_{ii}$ requires the inverse of $\bar{\mathbf{\Lambda}}_k$, which can be computed in practically linear time using conjugate gradient techniques [6]. Motivated by [5], these covariances can be efficiently approximated in constant time from their Markov blankets. Note also that Eq. (3.14) does not take into account the cross correlation between poses in the Mahalanobis metric, but this can be done with no substantial extra effort. The only difference is that instead of computing individual Markov blankets for each pose, the combined Markov blanket is used.

The average covariance is used to accommodate for the varying levels of estimation uncertainty both on the pose prior being evaluated, and on the past pose being compared. In case of a normal distribution, the Mahalanobis distance follows χ^2_{n-1}, i.e. a Chi-squared distribution with $n - 1$ degrees of freedom, with n the dimensions of one robot pose vector.

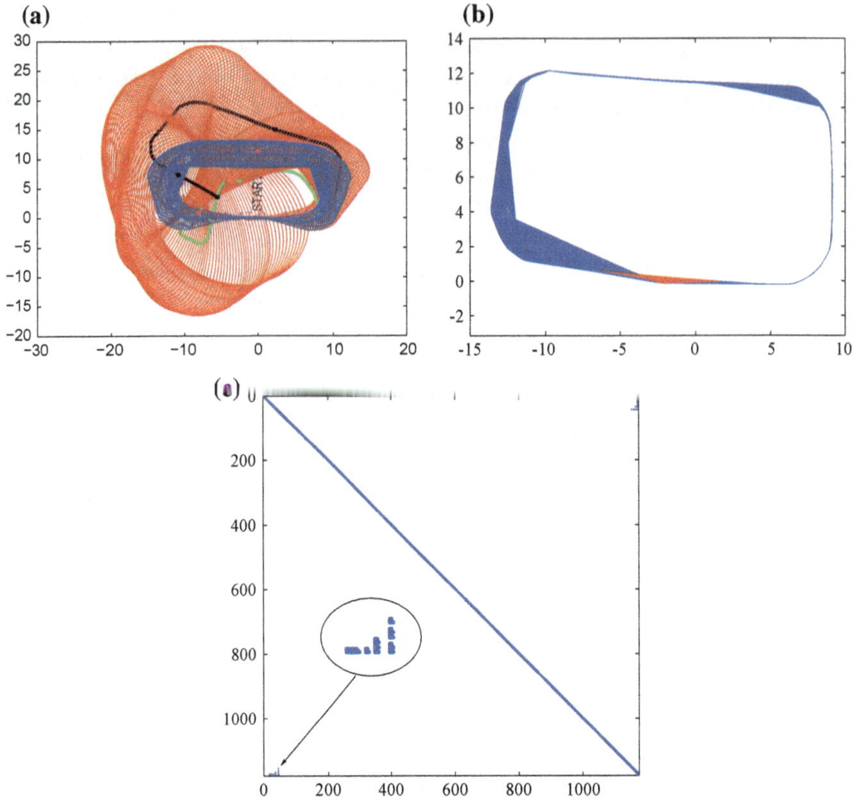

Fig. 3.1 A mobile robot has performed a loop trajectory. **a** Prior to adding the information relevant to the loop, a match hypothesis must be confirmed. If asserted, we could change the overall uncertainty in the trajectory from the red hyperellipsoids to the ones in *blue*. **b** If the test is made using conventional statistical tools, such as the Mahalanobis test, all the possible data association links indicated in blue should be verified. With a test based in information content just the links indicated in *red* should be verified. **c** A loop closure event adds only a few non-zero off-diagonal elements to the sparse information matrix (see zoomed region)

Many nearby poses will satisfy this condition, as shown in Fig. 3.1b. At the start of a SLAM run, when covariances are small, only links connecting very close poses will satisfy the test. But, as error accumulates, pose covariances grow covering larger and larger areas of matching candidates.

Due to linearization effects, adding information links for all possible matches produces overconfident estimates that in the long run lead to filter inconsistency. Thus, our update procedure must pass a second test. The aim of this second test is to allow updating using only links with high informative load. This happens when a pose with a large covariance can be linked with a pose with a small uncertainty.

$$d_B = \frac{1}{2} ln \frac{\left| \frac{\boldsymbol{\Sigma}_{kk} + \boldsymbol{\Sigma}_{ii}}{2} \right|}{\sqrt{| \boldsymbol{\Sigma}_{kk} || \boldsymbol{\Sigma}_{ii} |}} \tag{3.15}$$

The above expression refers to the second term of the Bhattacharyya distance, and gives a measure of separability in terms of covariance difference [20]. This test is typically used to discern between distinct classes with close means but varying covariances. Given that the value of d_B increases as the two covariances $\boldsymbol{\Sigma}_{kk}$ and $\boldsymbol{\Sigma}_{ii}$ are more different from each other. The Bhattacharyya covariance separability measure is symmetric, and we need to test whether the current pose covariance is larger than the i-th pose it is being compared with. This is done by analyzing the area of uncertainty of each estimate comparing the determinants of $\boldsymbol{\Sigma}_{kk}$ and $\boldsymbol{\Sigma}_{ii}$. The reason is that we only want to update the overall estimate with information links to states that had smaller uncertainty than the current state. Figure 3.1b shows in red the remaining links after the second test.

In a second phase we still must certify the presence of a loop closure event to update the entire pose estimate and to reduce overall uncertainty. When sensor registration can be established, the computed pose constraint is used in a one-step update of the information filter, as shown in Eqs. (3.12) and (3.11). A one-step update in information form changes the entire history of poses adding a constant number of non-zero off-diagonal elements in the information matrix as shown in the Fig. 3.1c. This sparsity can be controlled by reducing the confidence on sensor registration when testing for a loop-closure event.

3.1.3.2 Joint Measure of Information Content Between Poses

This strategy first selects loop closing poses by measuring its distance with respect to the last robot pose. This is done by computing relative displacement in belief space, d, from the robot pose, x_k, to any other previous pose in the trajectory, x_i, which can be can be estimated as a Gaussian with parameters [11]

$$\mu_d = h(\mu_k, \mu_i), \quad \text{and} \tag{3.16}$$

$$\boldsymbol{\Sigma}_d = [\mathbf{H}_i \ \mathbf{H}_k] \begin{bmatrix} \boldsymbol{\Sigma}_{ii} & \boldsymbol{\Sigma}_{ik} \\ \boldsymbol{\Sigma}_{ik}^\top & \boldsymbol{\Sigma}_{kk} \end{bmatrix} [\mathbf{H}_i \ \mathbf{H}_k]^\top , \tag{3.17}$$

where $\mathbf{H}_k$ and $\mathbf{H}_i$ are the Jacobians of h with respect to poses k and i, evaluated at the state means μ_k and μ_i, respectively, and $\boldsymbol{\Sigma}_{ki}$ is the cross correlation between these two poses.

Marginalizing the distribution of the displacement, d, along each one of its dimensions, t, we get a one-dimensional Gaussian distribution $\mathcal{N}(\mu_t, \sigma_t^2)$ that allows to compute the probability of pose x_i being closer than v_t to pose x_k along such dimension

$$\begin{aligned} p_t &= \int_{-v_t}^{+v_t} \mathcal{N}(\mu_t, \sigma_t^2) \\ &= \frac{1}{2}\left[\operatorname{erf}\left(\frac{v_t - \mu_t}{\sigma_t \sqrt{2}}\right) - \operatorname{erf}\left(\frac{-v_t - \mu_t}{\sigma_t \sqrt{2}}\right)\right]. \end{aligned} \tag{3.18}$$

If for all dimensions, p_t is above a given threshold s, then configuration x_i is considered close enough to configuration x_k.

Thresholds v_t are defined from the sensor characteristics, e.g., the field of view of cameras or maximum distance for the laser scan alignment. The value for the threshold s can be adjusted with the uncertainty in the robot pose, decreasing it as the robot gets lost, however, so as to avoid to increase complexity in Pose SLAM this is usually fixed. Moreover, a further improvement can be added by organizing the poses in a tree, which is done in [11], where the computation of the set of neighbors for each pose during mapping is performed in logarithmic time.

Next, for those candidate poses we evaluate its information content before attempting to close a loop. The mutual information gain for a candidate link measures the amount of uncertainty removed from the state when the link is integrated into the filter.

For Gaussian distributions, it is given by the logarithm of the ratio of determinants of prior and posterior state covariances [21–24]. These determinants are proportional to the volume of the covariance hyper-ellipsoids of equiprobability. Thus, this ratio is related with the number of times the state uncertainty shrinks once a loop is asserted.

As the covariance matrix is the inverse of the information matrix and taking into account Eq. (3.13) we have that the mutual information gain of a candidate link between poses i and k is

$$\mathcal{I} = \frac{1}{2} \ln \frac{|\boldsymbol{\Lambda} + \Delta\boldsymbol{\Lambda}|}{|\boldsymbol{\Lambda}|}. \tag{3.19}$$

Taking the natural logarithm, this measure is expressed in nats, and it can be evaluated efficiently as [11]

$$\mathcal{I}_{ki} = \frac{1}{2} \ln \frac{|\mathbf{S}_{ki}|}{|\boldsymbol{\Sigma}_z|}, \tag{3.20}$$

where $\boldsymbol{\Sigma}_z$ is the sensor registration error, $\mathbf{S}_{ki}$ is the innovation covariance

$$\mathbf{S}_{ki} = \boldsymbol{\Sigma}_z + [\mathbf{H}_k\ \mathbf{H}_i] \begin{bmatrix} \boldsymbol{\Sigma}_{kk} & \boldsymbol{\Sigma}_{ki} \\ \boldsymbol{\Sigma}_{ki}^\top & \boldsymbol{\Sigma}_{ii} \end{bmatrix} [\mathbf{H}_k\ \mathbf{H}_i]^\top. \tag{3.21}$$

If the result is above a given threshold, γ, sensor registration is needed to assert data association. When the real sensor covariance is computed during sensor registration (e.g. using the technique introduced in Chap. 2), it can be used to recompute the gain measure to ultimately decide whether or not to update the state with the new link. In the rest of the methods reported in this book we use this data association approach as it considers the effect of the candidate link information gain on the whole state.

3.1.4 State Sparsity

All delayed-state SLAM approaches, including Pose SLAM, accumulate robot poses over time, increasing the size of the state. One alternative to maintain a state vector of manageable size is to marginalize out redundant poses from the state. However, this is computationally costly and it degrades the sparsity of the information matrix. Since state marginalization is expensive, an alternative is to approximate pose marginalization using local Chow-Liu trees to keep sparsity [25]. In contrast, to overcome this issue, in Pose SLAM [11] it is preferred to add only non redundant poses and highly informative links to the state.

A new pose x_k is considered redundant when it is too close to any pose x_i, already in the trajectory. That is, if for all dimensions, p_t, computed with Eq. (3.18), is above a given threshold s, then pose x_k is considered close to the robot pose x_i and it is considered redundant. If no pose is close to x_k, it is included in the state. However, if the new pose allows to establish an informative link, both the link and the pose are added to the map. Information content is evaluated with Eq. (3.20).

This strategy reduces the state size in Pose SLAM, restricting the representation to environment size, which is fixed, and not by the length of the trajectory, which can grow without bound.

3.2 6 DOF Pose SLAM

An implementation of Pose SLAM to deal with poses in SE(3) requires to define the corresponding motion and observation models that capture the relative motion in 6 DOF. Additionally, the parameterizations of rotations require some modifications to the basic operations of Pose SLAM.

3.2.1 Euler Angles Parameterization

We adopt the ZYX-Euler angles parametrization [26], where ϕ, θ, and ψ are the angles of rotation about the x, y, and z axis, respectively, which define the following rotation matrix

$$
\mathbf{R} = \mathrm{rot}(\phi, \theta, \psi)
$$

$$
= \begin{bmatrix} \mathbf{c}\theta\mathbf{c}\psi & \mathbf{s}\phi\mathbf{s}\theta\mathbf{c}\psi - \mathbf{c}\phi\mathbf{s}\psi & \mathbf{c}\phi\mathbf{s}\theta\mathbf{c}\psi + \mathbf{s}\phi\mathbf{s}\psi \\ \mathbf{c}\theta\mathbf{s}\psi & \mathbf{s}\phi\mathbf{s}\theta\mathbf{s}\psi + \mathbf{c}\phi\mathbf{c}\psi & \mathbf{c}\phi\mathbf{s}\theta\mathbf{s}\psi - \mathbf{s}\phi\mathbf{c}\psi \\ -\mathbf{s}\theta & \mathbf{s}\phi\mathbf{c}\theta & \mathbf{c}\phi\mathbf{c}\theta \end{bmatrix}. \tag{3.22}
$$

where $\mathbf{s}x$ and $\mathbf{c}x$ represent $sin(x)$ and $cos(x)$, respectively.

Given $\mathbf{R}$, the Euler angles can be recovered as follows

$$
\begin{aligned}
\psi &= \mathrm{yaw}\,(\mathbf{R}) &&= \mathrm{atan2}\left(r_{2,1}, r_{1,1}\right) \\
\theta &= \mathrm{pitch}\,(\mathbf{R}) &&= \mathrm{atan2}\left(-r_{3,1}, r_{1,1}\mathbf{c}\,(\psi) + r_{2,1}\mathbf{s}\,(\psi)\right) \\
\phi &= \mathrm{roll}\,(\mathbf{R}) &&= \mathrm{atan2}\left(r_{1,3}\mathbf{s}\,(\psi) - r_{2,3}\mathbf{c}\,(\psi), -r_{1,2}\mathbf{s}\,(\psi) - r_{2,2}\mathbf{c}\,(\psi)\right),
\end{aligned}
$$

where

$$
\mathbf{R} = \begin{bmatrix} r_{1,1} & r_{1,2} & r_{1,3} \\ r_{2,1} & r_{2,2} & r_{2,3} \\ r_{3,1} & r_{3,2} & r_{3,3} \end{bmatrix}.
$$

With this parametrization to represent rotations, we define a robot pose (the i-th component of the state vector $\mathbf{x}_k$) as follows

$$
x_i = \left[\mathbf{t}_i^\top, \, \mathbf{\Theta}_i^\top\right]^\top, \tag{3.23}
$$

where $\mathbf{t}_i = \left[x^{(i)}, y^{(i)}, z^{(i)}\right]^\top$ is the position of the robot and $\mathbf{\Theta}_i = \left[\phi^{(i)}, \theta^{(i)}, \psi^{(i)}\right]^\top$ its orientation.

Next, we define both the motion and observation models by extrapolating the compounding operation defined in [16] to poses in SE(3). The noise-free motion model from robot pose x_{k-1} to x_k is given by

$$
\begin{aligned}
x_k &= x_{k-1} \oplus u_k \\
&= \begin{bmatrix} \mathbf{t}_{k-1} + \mathbf{R}_{k-1}\,\Delta\mathbf{t}_k \\ \mathrm{roll}\,(\,\mathbf{R}_{k-1}\,\mathbf{R}_u\,) \\ \mathrm{pitch}\,(\,\mathbf{R}_{k-1}\,\mathbf{R}_u\,) \\ \mathrm{yaw}\,(\,\mathbf{R}_{k-1}\,\mathbf{R}_u\,) \end{bmatrix},
\end{aligned} \tag{3.24}
$$

where $u_k = \left[\Delta\mathbf{t}_k^\top, \, \Delta\mathbf{\Theta}_k^\top\right]^\top$ is the relative motion between x_k and x_{k-1}, w.r.t. x_{k-1}, produced by the current motion command, with $\Delta\mathbf{\Theta}_k = \left[\Delta\phi^{(k)}, \Delta\theta^{(k)}, \Delta\psi^{(k)}\right]^\top$,

$$
\mathbf{R}_{k-1} = \mathrm{rot}\,(\phi^{(k-1)}, \theta^{(k-1)}, \psi^{(k-1)}),
$$

and

$$
\mathbf{R}_u = \mathrm{rot}\,(\Delta\phi^{(k)}, \Delta\theta^{(k)}, \Delta\psi^{(k)}).
$$

The noise-free observation model, which indicates the relative motion between any robot pose x_i and the current robot pose x_k, is given by

$$
\begin{aligned}
z_{ki} &= h(x_k, x_i) \\
&= \ominus x_k \oplus x_i \\
&= \begin{bmatrix} \mathbf{R}_k \left[\mathbf{t}_i - \mathbf{t}_k\right] \\ \text{roll}\,(\,\mathbf{R}_k^\top\ \mathbf{R}_i\,) \\ \text{pitch}\,(\,\mathbf{R}_k^\top\ \mathbf{R}_i\,) \\ \text{yaw}\,(\,\mathbf{R}_k^\top\ \mathbf{R}_i\,) \end{bmatrix},
\end{aligned}
\tag{3.25}
$$

where z_{ki} is measured w.r.t. x_k, with

$$\mathbf{R}_k = \text{rot}\,(\phi^{(k)}, \theta^{(k)}, \psi^{(k)})$$

and

$$\mathbf{R}_i = \text{rot}\,(\phi^{(i)}, \theta^{(i)}, \psi^{(i)}).$$

For this angle parameterization, once having computed the Jacobians for both the motion and observation models, the set of equations presented in Sect. 3.1 can be used without any further modification.

3.2.2 *Quaternion Parameterization*

One drawback of the parameterization with Euler angles is the loss of one degree of freedom or gimbal lock. An alternative to this issue is to employ a quaternion parameterization. For this parameterization, we define one robot pose (the i-th component of the state vector $\mathbf{x}_k$) as follows

$$x_i = \left[\mathbf{t}_i{}^\top\ \mathbf{q}_i{}^\top\right]^\top, \tag{3.26}$$

where $\mathbf{t}_i$ indicates the position of the robot and $\mathbf{q}_i$ is a unit norm quaternion expressing the robot orientation.

The noise-free motion model is defined using an extrapolation of the compounding operation [16] for quaternions as we show next,

$$
\begin{aligned}
x_{k+1} &= f(x_k, u_k) \\
&= x_k \oplus u_k \\
&= \begin{bmatrix} \tilde{\mathbf{t}}_k + \mathbf{q}_k \otimes \Delta\tilde{\mathbf{t}}_k \otimes \mathbf{q}_k^{-1} \\ \mathbf{q}_k \otimes \Delta\mathbf{q}_k \end{bmatrix},
\end{aligned}
\tag{3.27}
$$

where the notation $\tilde{\mathbf{a}}$ means an ordinary vector $\mathbf{a} = [x, y, z]^\top$ in a 3-dimensional space with the last element equal to zero, i.e. $\tilde{\mathbf{a}} = [x, y, z, 0]^\top$, and the operator $\otimes$ indicates a quaternion multiplication. Additionally, the relative motion given by the odometry data is represented with u_k, which is given by the relative traveled distance $\Delta\mathbf{t}_k$ and the relative rotation change $\Delta\mathbf{q}_k$.

Note that the translation part in Eq. (3.27) gives a quaternion with a scalar part equal to zero; however, in x_{k+1} this zero is omitted.

We form the noise-free observation model also using the compounding operations. The noise-free measurement model is given by Eq. (3.28), which tells us how much the robot has moved between the current pose x_k and any robot pose x_i, w.r.t x_k,

$$\begin{aligned} z_{ki} &= h(x_k, x_i) \\ &= \ominus x_k \oplus x_i \\ &= \begin{bmatrix} \mathbf{q}_i^{-1} \otimes (\tilde{\mathbf{t}}_k - \tilde{\mathbf{t}}_i) \otimes \mathbf{q}_i \\ \mathbf{q}_i^{-1} \otimes \mathbf{q}_k \end{bmatrix}. \end{aligned} \tag{3.28}$$

Note that in this expression the translation also gives a quaternion with a scalar part equal to zero, which in z_{ki} is also omitted.

Before applying the formulation introduced in Sect. 3.1 we need to compute the Jacobians for both the motion and observation models. Additionally, for this representation of rotations we need to consider the following.

In the state augmentation (Eqs. 3.6–3.7) and state update (Eqs. 3.11–3.12) operations we need to invert $\mathbf{\Sigma}_u$ and $\mathbf{\Sigma}_z$, that is, the motion and measurement noise covariances, respectively. However, one issue with the quaternion parameterization is that the ensuing covariance matrices are rank deficient by one due to the normalized quaternion representation.

To solve this problem we compute the pseudo-inverse matrices $\mathbf{\Sigma}_u^+$ and $\mathbf{\Sigma}_z^+$, for the motion and measurement noise covariances, respectively. To this end, assuming $\mathbf{\Sigma}_u$ and $\mathbf{\Sigma}_z$ to be block diagonal matrices, with each block matrix corresponding to the translation and rotation variables, a pseudo-inverse matrix should be computed for the block that contains the quaternion part.

In order to obtain $\mathbf{\Sigma}_u^+$, we assume the rotation part $\mathbf{Q}_e$ of the motion noise covariance to be initially specified in Euler angles. Then, our first step is to transform this covariance to be expressed in quaternions. This is done with a first-order linear propagation of the noise in Euler angles. To do this, we define the function

$$g : \mathcal{X}_e \rightarrow \mathcal{X}_q, \tag{3.29}$$

which transforms robot orientations in Euler angles $\mathcal{X}_e \subseteq \mathrm{SO}(3)$ to robot orientations in quaternions $\mathcal{X}_q$, whose first-order linearization, about the orientation mean $\boldsymbol{\mu}_e$ of the current robot orientation x_e in Euler angles, is given by

$$g(x_e) \approx g(\boldsymbol{\mu}_e) + \mathbf{G}(x_e - \boldsymbol{\mu}_e), \tag{3.30}$$

where

$$\mathbf{G} = \left.\frac{\partial g}{\partial \mathbf{x}_e}\right|_{\boldsymbol{\mu}_e}. \tag{3.31}$$

Thus, the covariance of the rotation noise in quaternions is given by

$$\mathbf{Q}_q = \mathbf{G}\mathbf{Q}_e\mathbf{G}^\top, \tag{3.32}$$

which is rank deficient by one because of the normalized quaternion representation.

Next, we compute the pseudo-inverse for $\mathbf{Q}_q$ as follows

$$\mathbf{Q}_q^+ = \mathbf{G}(\mathbf{G}^\top\mathbf{G}\mathbf{Q}_e\mathbf{G}^\top\mathbf{G})^{-1}\mathbf{G}^\top. \tag{3.33}$$

Finally, the pseudo-inverse of the motion noise is given by

$$\boldsymbol{\Sigma}_u^+ = \begin{bmatrix} \mathbf{Q}_t^{-1} & \mathbf{0} \\ \mathbf{0} & \mathbf{Q}_q^+ \end{bmatrix}, \tag{3.34}$$

where $\mathbf{Q}_t$ represents the translational components of the motion noise covariance $\boldsymbol{\Sigma}_u$.

Next, applying the aforementioned process to find a pseudo-inverse for the observation noise we obtain

$$\boldsymbol{\Sigma}_z^+ = \begin{bmatrix} \mathbf{V}_t^{-1} & \mathbf{0} \\ \mathbf{0} & \mathbf{G}(\mathbf{G}^\top\mathbf{G}\mathbf{V}_e\mathbf{G}^\top\mathbf{G})^{-1}\mathbf{G}^\top \end{bmatrix}, \tag{3.35}$$

with $\mathbf{V}_t$ the translational components of the measurement noise covariance $\boldsymbol{\Sigma}_z$ and $\mathbf{V}_e$ its rotational components in Euler angles.

Finally, a further requirement for the Pose SLAM algorithm is to enforce quaternion normalization at each iteration. Thus, for the rotation part $x_q \sim \mathcal{N}(\boldsymbol{\mu}_q, \boldsymbol{\Sigma}_q)$ of the robot pose estimate, we enforce quaternion normalization as follows

$$\boldsymbol{\mu}_q' = g_k(\boldsymbol{\mu}_q) \tag{3.36}$$

and

$$\boldsymbol{\Sigma}_q' = \mathbf{G}_k\boldsymbol{\Sigma}_q\mathbf{G}_k^\top, \tag{3.37}$$

where g_k is the function that performs quaternion normalization, that is,

$$g_k(\mathbf{q}) = \frac{\mathbf{q}}{\parallel \mathbf{q} \parallel},$$

with Jacobian

$$\mathbf{G}_k = \left.\frac{\partial g_k}{\partial \mathbf{q}}\right|_{\boldsymbol{\mu}_q}.$$

3.3 Traversability Map Building

This method takes as input an occupancy grid map, with each cell representing the presence of an obstacle at that location in the environment, and transforms it into another grid map whose cells indicate the maximum linear velocity that guarantees a collision-free path for a specific mobile robot. To this end, it considers the kinematic model of the mobile robot, thus giving a tailored map for each different mobile platform, which is useful for the navigation of a heterogeneous fleet of robots moving in the same environment.

We consider the configuration space of the robot to be discretized in position according to the resolution of the grid map and in orientation by a desired resolution. Thus, given the action space $\mathcal{A} = \mathcal{V} \times \Omega$, with $\mathcal{V}$ and Ω the sets of all possible linear and angular velocities, respectively. We compute the set $\mathcal{V}(x, y, \theta)$ of all linear velocities that generate a collision-free path departing from robot configuration $[x, y, \theta]^\top$, with (x, y) the cell's position and θ the robot's orientation. To this end, for a given a cell, for each robot's orientation and every control action (v_j, ω_j), with $\mathrm{v}_j \in \mathcal{V}$ and $\omega_j \in \Omega$, using the kinematic model of our mobile robot, we generate a path by iterating a fixed period of time. Then, we add the linear velocity v_j to $\mathcal{V}(x, y, \theta)$ if the resulting path is within the free space, wherein collision detection is performed using input gridmap.

Then, we compute the set defined by the union of the maximum linear velocities to scape from cell (x, y) with orientation θ that lead to a collision free trajectory, that is, we compute

$$\mathcal{V}_m(x, y) = \bigcup_{\theta \in \Theta} max(\mathcal{V}(x, y, \theta)), \tag{3.38}$$

which, for a given cell's position (x, y), contains the maximum linear velocities for all orientations Θ.

Thus, the traversability map is defined by the function that associates every cell position (x, y) to the maximum linear velocity v_f that guarantees a collision-free path

$$m : (x, y) \rightarrow \mathrm{v}_f, \tag{3.39}$$

where

$$\mathrm{v}_f = min(\mathcal{V}_m(x, y)). \tag{3.40}$$

We take the minimum in Eq. (3.40) to make sure that the robot does not collide with an obstacle after departing from all the considered orientations. As $\mathcal{V}_m$ contains the maximum velocities departing from different orientations, certainly, taking the maximum of this would lead to a collision free trajectory for their corresponding initial orientation, from where it was selected, however, we cannot guarantee that the robot does not collide with an obstacle if we apply the same velocity but starting from a different initial orientation.

3.4 Mapping with Pose SLAM

Next, we show three maps computed with Pose SLAM. The first shows the application of the 2D Pose SLAM implementation using as inputs relative measurements computed with our visual odometry approach presented in Chap. 2. The second map shows a 3D map built by our 6 DOF Pose SLAM implementation with three dimensional laser range scans as the main input data, where the relative measurements were computed with a hierarchical ICP implementation [27]. The third result is a traversability map computed by post-processing the 3D map.

3.4.1 *Visual Odometry Map*

For this experiment, we collected dead-reckoning readings and stereo images with a Segway RMP 400 robotic platform, equipped with two PointGrey Flea2 cameras for about 350 m in an outdoor environment.

The Segway dead reckoning readings and the visual pose constraints are modelled with noise covariances $\mathbf{\Sigma}_u = \text{diag}(0.0316\,\text{m}, 0.0158\,\text{m}, 0.1104\,\text{rad})^2$, and $\mathbf{\Sigma}_z = \text{diag}(0.2\,\text{m}, 0.2\,\text{m}, 0.03\,\text{rad})^2$, respectively, and the uncertainty of the initial pose is set to $\mathbf{\Sigma}_0 = \text{diag}(0.1\,\text{m}, 0.1\,\text{m}, 0.09\,\text{rad})^2$. Note that the static motion and measurement covariances are chosen to overestimate the true covariances.

Experimentally, we observed that images taken in poses farther away than ± 4.5 m in x, ± 4.5 m in y or ± 1.04 rad in orientation can not be safely matched and, consequently, those are the thresholds in v_t used to detect nearby poses, with $s = 0.1$. In this experiment, the minimum information gain is set to $\gamma = 1.5$ nats.

Figure 3.2 shows the trajectory estimated by Pose SLAM. The red dots and lines represent the trajectory estimated fusing encoder and visual odometry, and the green lines indicate loop closure constraints established by registering stereo images at non-consecutive poses. Although vision-based pose constraints can fail in translation estimation, they provide quite accurate rotation estimation, and it helped to correct the raw odometry information, which is especially poor when the vehicle turns.

3.4.2 *3D Volumetric Map*

Next, we show mapping results with 6 DOF Pose SLAM, using three dimensional laser range scans as the main input data. The datasets were acquired in an outdoor urban environment. The experimental site was the Barcelona Robot Lab, located at the Campus Nord of the Universitat Politècnica de Catalunya. This experimental area has over $15,000\,\text{m}^2$, several levels and underpasses, poor GPS coverage, moderate vegetation, several points with aliasing, large amounts of regularity from building structures, and sunlight exposure severely subject to shadows. The robot was tele-

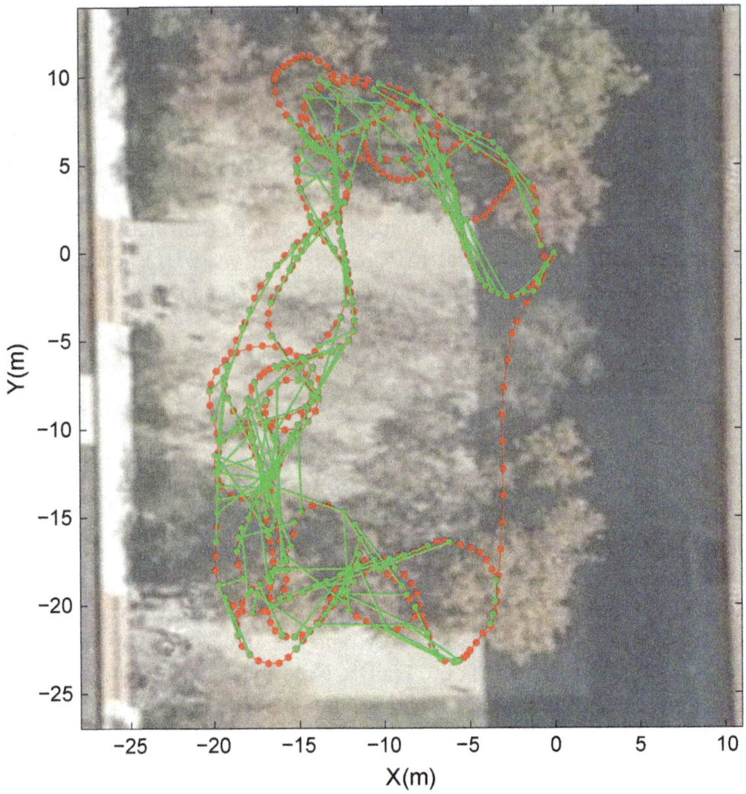

Fig. 3.2 Pose SLAM map built with encoder odometry and stereo vision data

operated through this site along a path of over 600 m (see Fig. 3.4a). Along the way the robot acquired 3D Laser Data with an IRI's proprietary 3D scanning system, installed atop an Activmedia Pioneer 2AT robotic platform. The system yields 3D point clouds with ranges up to 30 m, and sizes of about 76,000 points. The sensor noise level is ±5 cm in depth estimation for each laser beam. Figure 3.3 portrays the complete device.

The relative motion measurements were computed with the optimized ICP implementation shown in [27], which employs a hierarchical structure that uses a point-to-plane error metric at the coarsest level and a point-to-point metric at finer levels, and that weights differently rotations and translations. These relative motion measurements were introduced to the Pose SLAM algorithm using Euler angles. Figure 3.4a contains results from state augmentation purely from concatenation of ICP computed motion constraints. The hyper-ellipsoids shown indicate marginal covariances of the robot position. This open loop traverse causes an increment of the accumulated estimation error. The mapping strategy discussed closes 19 loops, with the consequent improvement on localization uncertainty, as depicted in Fig. 3.4b. The complete alignment of the 3D point clouds is shown in Fig. 3.5.

Fig. 3.3 3D laser range finder mounted on our robotic platform

Sensor covariance was approximated with a first order error propagation by computing the implicit function Jacobian for ICP's point-to-point unconstrained minimization as shown in [28, 29]. Although it is formulated only for the point-to-point error metric, whilst the ICP implementation is optimized for performance with a hierarchical structure that uses a point-to-plane error metric at the coarsest level and a point-to-point metric at finer levels, in our experiments we obtain a map qualitatively consistent with geo-referenced CAD model of the same environment [30]. We hypothesize that the overall estimation error of our results varies from 5 cm to

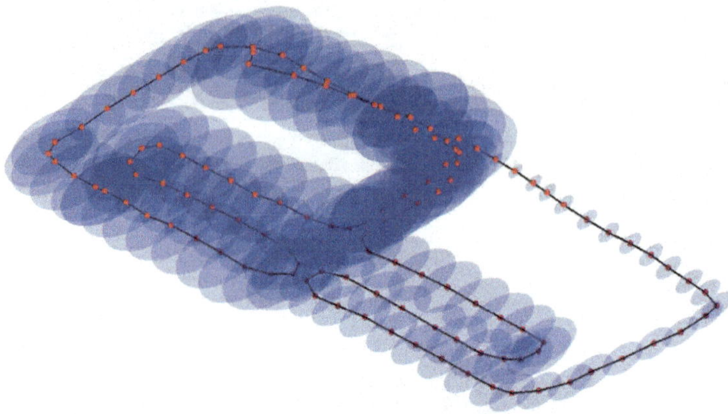

(a) State estimate before loop closures.

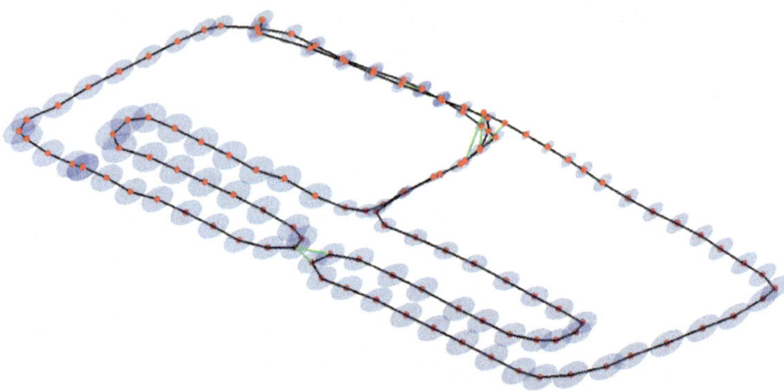

(b) Estimated trajectory after loop closures.

Fig. 3.4 6D range-based SLAM results

50 cm. These values however cannot be verified with sufficient precision since no sufficiently accurate ground truth is available.

3.4.3 Traversability Map

Given the aforementioned 3D map (Fig. 3.5) we computed the traversability map shown in Fig. 3.6 for the robot depicted in Fig. 3.3 using the approach introduced in Sect. 3.3. To do this, we discretized the configuration space in 10 cm for the robot position and 0.25 rad for robot orientation, and the action space in 0.1 m/s for linear velocity and 0.01 rad/s for angular velocity. From Fig. 3.6, we can note that, as expected, the highest speeds can be achieved in the open areas of the environment, while narrower regions received the lowest values for the maximum speed.

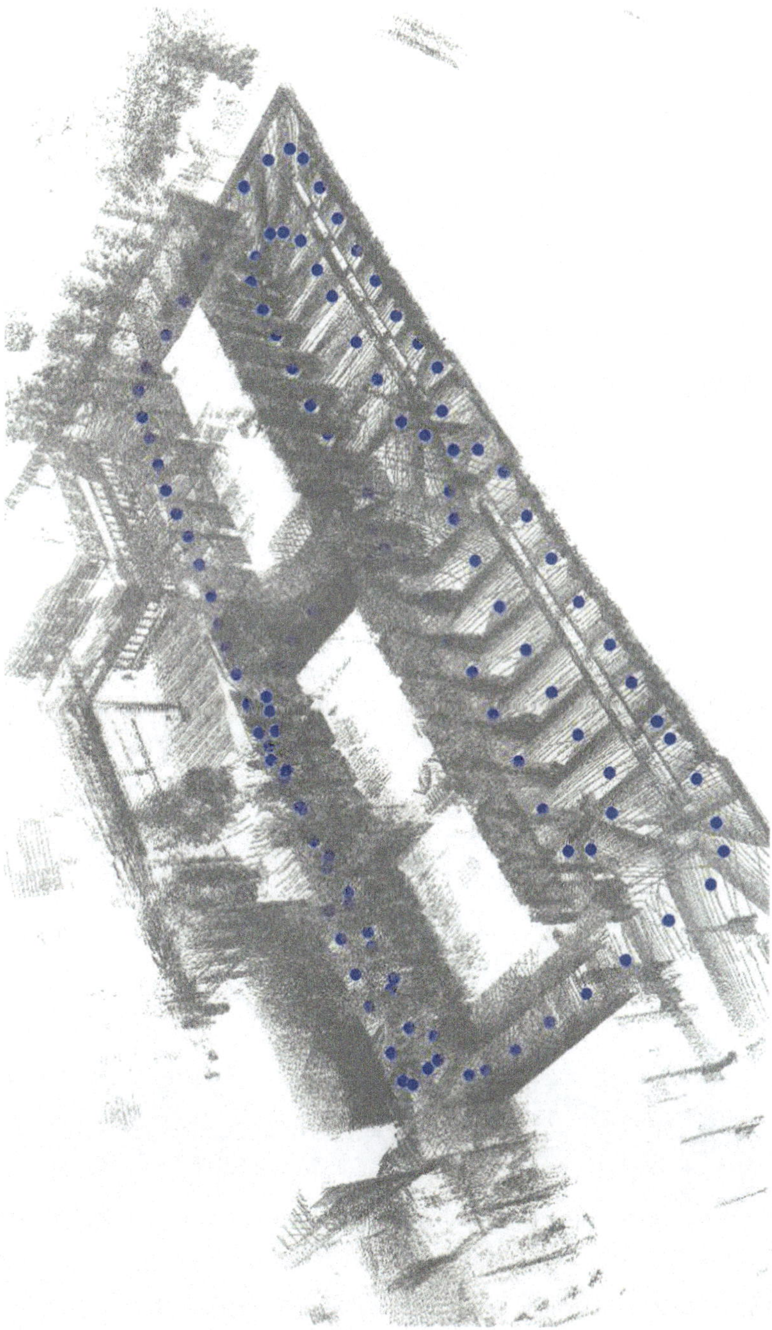

Fig. 3.5 Final 3D map of the experimental site computed with our approach

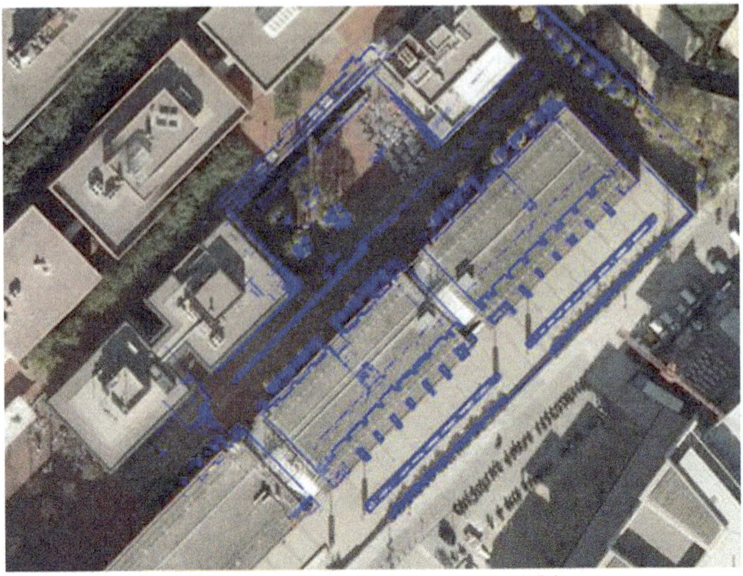

(a) 2D Layer superimposed on an aerial image

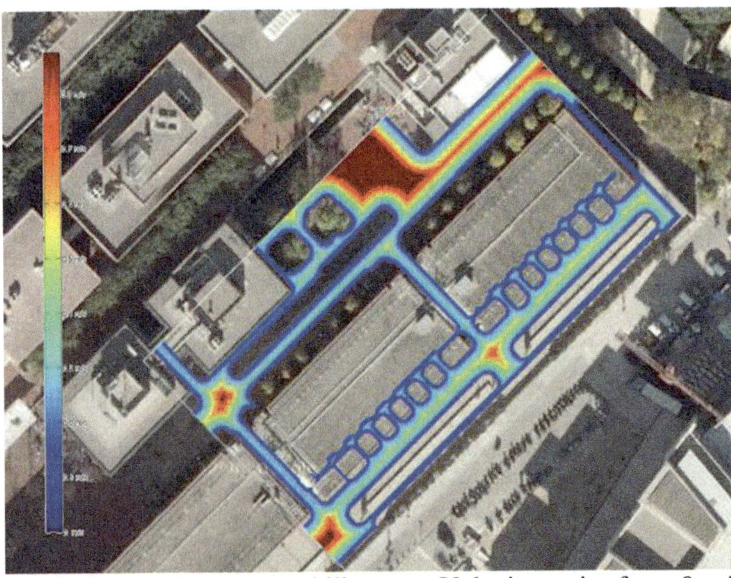

(b) Corresponding traversability map. Velocity varies from 0 m/s (blue) to 1 m/s (red).

Fig. 3.6 Traversability map from 2D layers of the aligned 3D point clouds

3.5 Bibliographical Notes

The problem of building a map of an unknown environment with a mobile robot, while at the same time being localized relative to this map was introduced by a series of seminal papers [2, 31–33] during the second half of the 1980s. It was initially referred to as Concurrent Map Building or CML [34] and later known as Simultaneous Localization and Mapping or SLAM [35, 36]. Eventually, this problem was extensively studied by many subsequent works and their evolution has even been reviewed in different surveys that have appeared along the way, such as [37–39].

Soltutions to SLAM have evolved in different directions, mainly driven by the challenges that this problem imposes. One of these problems is related to the precision issues derived by the state estimation approach employed to solve SLAM. The problem of SLAM has been addressed as a state estimation problem in which perception and motion uncertainties are coupled. Traditionally, the problem has been solved with Extended Kalman filtering [1, 4], a solution usually referred to as EKF-SLAM. The technique has allowed to identify the basic properties of such a coupled state estimation problem [3], however, it presents drawbacks with respect to precision due to linearizations and scalability. The linearization of the perceptual models introduces optimistic estimations in every iteration, which in the limit produce filter inconsistency. This is specially noticed in the case of rotations [40, 41]. To mitigate the effect of linearization other estimation techniques can be used, such as the unscented Kalman filter (UKF) [42, 43], information filters [5, 6], or particle filters [9].

Scalability is also a concern in landmark-based SLAM, the robot pose and the map of features has quadratic computational complexity, limiting the approach to relatively small environments. This computational cost can be alleviated using the Extended Information Filter (EIF) and its alternative parametrization of Gaussian distributions based on the information vector and the information matrix. The information matrix in landmark-based SLAM is approximately sparse with very small matrix entries for distant landmarks [5]. These entries can be removed, compacting the map and speeding up the filter. If instead of only estimating the last robot pose, the whole robot trajectory is included in the state together with the landmarks (an approach typically referred to as full SLAM [8, 9, 44]) a sparse information matrix is obtained without using approximations. Going one step further, in Pose SLAM [6, 7, 12, 45]. only the trajectory of the robot is included in the state and the landmarks are only used to derive relative motion constraints between poses. The result is an exactly sparse information matrix which grows with the number of poses and that only has non-null entries for those poses directly related by an observation.

The problem with a naive implementation of Pose SLAM is that the state grows indefinitely, independent of the size of the environment. Heuristic strategies can be found in the Pose SLAM literature to either reduce the size of the state representation by keeping only one pose every few meters [12, 46] or to restrict the number of links to a number linearly dependent on the number of poses [6]. In contrast, this problem has been solved with principled information-based measures by either pruning the graph of poses [25] for the laser-based SLAM case or by controlling the number of poses added to filter [11] independent of the sensor modality.

Information-based approaches are perfectly suited for off-line maximum likelihood estimation. In this case, data association is usually taken for granted, joint marginals are not necessary, and the estimate only includes the state mean, which is iteratively approximated representing the relative displacement between poses as quadratic constraints. This formulation is usually referred to as Graph-based SLAM and was initally introduced by Lu and Milios in 1997 [45]. In the Graph SLAM paradigm the problem is usually divided in two parts: graph constructing (SLAM front-end) and graph optimization (SLAM back-end). Most of the Graph SLAM solutions are graph optimization approaches or SLAM back-ends, which assume a graph has been built beforehand by the SLAM front-end. A comprehensive introduction and review of such a SLAM formulation was presented in [47].

Since the pioneering work introduced by Lu and Milios, it took some time to make the Graph SLAM formulation widely used because of the cost of solving the error minimization problem. Following [45], a number improvements to the Graph SLAM formulation appeared, many of them used iterative optimization techniques such as gradient descent [48, 49] or relaxation [50] to build a map. In [51], Duckett et al. proposed the usage of Gauss-Seidel relaxation to minimize the error in the network of constraints. Frese et al. [52] introduced multi-level relaxation (MLR), a variant of Gauss-Seidel relaxation that applies the relaxation at different levels of resolution. In [53] it was suggested a gradient descent approach to optimize pose graphs, which was later extended by Grisetti et al. [54] to increase its convergence speed.

A further improvement to the Graph SLAM methods was possible by exploiting the sparsity of the information matrix. An efficient sparse factorization for SLAM known as Square Root SAM was introduced in Dellaert et al. [10], which performs smoothing by Cholesky factorization of the complete, exactly sparse information matrix in every step using the Levenberg-Marquardt algorithm. Likewise, Konolige et al. [55] introduced a solution using Cholesky factorization, including a continuable Levenberg-Marquardt algorithm and focuses on a fast setup of the information matrix. Similarly in [56] the authors shared similar ideas with these systems.

Aside from batch optimization approaches, incremental Graph SLAM techniques were also proposed. In such techniques the cost of batch optimization is alleviated by applying matrix factorization updates. Kaess et al. [8] introduced an iterative variant of the Square Root SAM approach known as iSAM, which performs incremental updates using Givens rotations, with periodic batch factorization and relinearization steps. A further extension to iSAM was introduced in [57] with the advantage that both reordering and relinearization can be performed incrementally in every step. Relative Graph SLAM formulations were also introduced. In [58] Klein and Murray proposed an approach that splits the pose tracking and mapping into two separate tasks, where the first tracks the pose of the camera relative to the map, and the second thread creates a map with a subset of the frames. Similarly, Eade and Drummond [59] presented an approach that builds a graph of statistically independent submaps using a monocular camera. Among other relative Graph SLAM approaches are the CI-Graph SLAM [60], Tectonic SAM [61] and Atlas [62]. Similarly, Sibley et al.

[63] introduced a method that couples the relative formulation with locally restricted optimization, operating on a manifold. Likewise,

Graph SLAM approaches have evolved to the point to be gradually faster and, as of now, these methods are the state-of-the-art approaches to compute the SLAM estimate in terms of speed and accuracy. However, such approaches alone do not solve the whole SLAM problem as they require a SLAM front-end to build the graph.

Data association is another issue in SLAM and a key component of a SLAM front-end. It is specially difficult in environments where landmarks are difficult to distinguish as well as in environments with aliasing. An important advance in the data association problem was the concept of batch gating, where multiple associations are considered simultaneously [40, 64]. Furthermore, the standard formulation of the EKF-SLAM solution is especially fragile to incorrect association of observations to landmarks [65], and in order to validate the history of data association, temporal landmark quality measures and a temporal landmark quality test were proposed in [66]. These quality measures allow longer mapping sequences through the elimination of inconsistent observations. This removal of weak landmarks from the state vector and state covariance matrix did not violate the convergence properties of SLAM. Other ways to perform reliable data association include other sensing modalities, such as vision. For instance, appearance signatures [18, 19] are useful to predict a possible association, such as closing a loop. Another alternative is to exploit the filter information to constrain the search for sensory matches only to few neighboring poses [6, 7, 11]. For consistent estimates, this latter option is more efficient and less affected by perceptual aliasing.

Currently, SLAM front-end algorithms alone cannot guarantee accuracy all the time, for instance, it might be possible that data association approaches fail leading to inconsistent estimates. Therefore, several techniques have been proposed to add robustness to the graph optimization process, which are usually referred to as robust Graph SLAM back-ends.

Most of the robust SLAM back-ends have focused on identifying and discarding spurious constraints. In [67, 68], it was introduced a switchable constraints approach where each loop closure is associated with an additional variable known as switch that controls the contributed residual error of the corresponding link. Thus, the optimizer has to find the optimal configuration of poses as well as the switchable variables. The effect of the switchable variables can be seen as scaling the information matrix with the square of the switch variable. This is also followed in [69], however, instead of introducing a switchable variable and letting the solver find the optimal values for it, it computes the switch values using the corresponding covariance for the switch prior and the current normalized residual. In [70], instead of scaling the information matrix, the authors employed a max-mixture distribution to include multiple hypotheses on the noise distribution of a measurement. Then, the distribution that best explains the loop closure is selected and employed in the optimization process. An alternative to scaling the covariance matrix or providing a fixed guess for its value was introduced in [71], where it was proposed an approach that instead carries out a series of statistical tests. This method consists of a consensus based algorithm that checks for loops that lead to successful convergence of the graph-SLAM problem. Another alternative was introduced in [72], where instead of dealing with the rejection of outliers during pose estimation, the authors proposed to select the best measurements instead. That is, such a method provides the set of reliable measurements, rather than pose estimates. A comprehensive comparative study of such robust Graph SLAM back-ends was introduced in [73, 74].

Another problem in SLAM is that of the environment representation. Early works in SLAM, usually in indoor settings, modeled the world as a set of two dimensional landmarks, however, this representation is of little help in more complex and unstructured environments. Diverse types of representation can be used. As noted in [75], according to the reference frame to which estimates are linked, we can distinguish between world-centric and sensor-centric representations. Moreover, if the world is described by a qualitative or geometrical description we can distinguish between topological and geometrical maps, respectively.

Geometrical representations in SLAM are the most popular. They can be further classified by the degrees-of-freedom (DOF) of the robot pose and the dimensionality of sensor data, as noted in [76], yielding four typical representations: planar maps with a 3 DOF SLAM, planar or slice-wise maps with a 6 DOF SLAM, 3D maps with a 3 DOF SLAM, and 3D maps with 6 DOF SLAM. This latter solution has received special attention. In [76] this problem is addressed by a scan alignment approach for 3D scans gathered by a rotating laser scanner sensor, where the alignment of two scans is done mainly by improvements to the basic ICP algorithm proposed in [77]. For the same type of solution and with a similar a sensor setting, a delayed-state framework with an Extended Kalman Filter was proposed in [78], and eventually a scalable version with only informative links and poses was presented in [30] using the Pose SLAM approach in 6 DOF.

The approaches so far discussed are passive in the sense that the robot only estimates the model of the environment, but without taking any decisions on its trajectory. Another challenge in SLAM is to compute the appropriate robot actions to reduce the uncertainty about its own localization and the map, while at the same time optimizing coverage [23, 30, 79, 80].

Many issues still remain to be solved, e.g. dynamic environments, multirobot SLAM, lifelong SLAM techniques. Another related issue, usually neglected in SLAM implementations, is to endow the robot the ability to actually use the maps it builds, just after it builds them, which might bring closer the possibility of long-term autonomous existence. This work contributes on this venue with a path planning under uncertainty using Pose SLAM maps. Chapter 4 will introduce this problem, including a bibliographical review on the related work to this issue as well. Furthermore, another contribution of this work is an active Pose SLAM approach, which will be introduced in Chap. 5, together with a review of the related work that will complement this section.

References

1. M.W.M.G. Dissanayake, P. Newman, S. Clark, H.F. Durrant-Whyte, M. Csorba, A solution to the simultaneous localization and map building (SLAM) problem. IEEE Trans. Robot. Autom. **17**(3), 229–241 (2001)
2. R.C. Smith, P. Cheeseman, On the representation and estimation of spatial uncertainty. Int. J. Robot. Res. **5**(4), 56–68 (1986)
3. J. Andrade-Cetto, A. Sanfeliu, The effects of partial observability when building fully correlated maps. IEEE Trans. Robot. **21**(4), 771–777 (2005)
4. J. Andrade-Cetto, A. Sanfeliu, *Environment Learning for Indoor Mobile Robots. A Stochastic State Estimation Approach to Simultaneous Localization and Map Building*, Springer Tracts in Advanced Robotics, vol. 23 (Springer, 2006)
5. S. Thrun, Y. Liu, D. Koller, A.Y. Ng, Z. Ghahramani, H. Durrant-Whyte, Simultaneous localization and mapping with sparse extended information filters. Int. J. Robot. Res. **23**(7–8), 693–716 (2004)
6. R.M. Eustice, H. Singh, J.J. Leonard, Exactly sparse delayed-state filters for view-based SLAM. IEEE Trans. Robot. **22**(6), 1100–1114 (2006)
7. V. Ila, J. Andrade-Cetto, R. Valencia, A. Sanfeliu, Vision-based loop closing for delayed state robot mapping, in *Proceedings of IEEE/RSJ International Conference on Intelligent Robots and Systems* (San Diego, Nov 2007), pp. 3892–3897
8. M. Kaess, A. Ranganathan, F. Dellaert, iSAM: Incremental smoothing and mapping. IEEE Trans. Robot. **24**(6), 1365–1378 (2008)
9. M. Montemerlo, S. Thrun, *FastSLAM: A Scalable Method for the Simultaneous Localization and Mapping Problem in Robotics*, Springer Tracts in Advanced Robotics, vol. 27 (Springer, 2007)
10. F. Dellaert, M. Kaess, Square root SAM: simultaneous localization and mapping via square root information smoothing. Int. J. Robot. Res. **25**(12), 1181–1204 (2006)
11. V. Ila, J.M. Porta, J. Andrade-Cetto, Information-based compact Pose SLAM. IEEE Trans. Robot. **26**(1), 78–93 (2010)
12. K. Konolige, M. Agrawal, FrameSLAM: from bundle adjustment to realtime visual mapping. IEEE Trans. Robot. **24**(5), 1066–1077 (2008)
13. S.J. Julier, J.K. Uhlmann, A counter example to the theory of simultaneous localization and map building, in *Proceedings of IEEE International Conference on Robotics and Automation* (Seoul, May 2001), pp. 4238–4243
14. T. Bailey, J. Nieto, J. Guivant, M. Stevens, E. Nebot, Consistency of the EKF-SLAM algorithm, in *Proceedings of IEEE/RSJ International Conference on Intelligent Robots and Systems* (Beijing, Oct 2006), pp. 3562–3568
15. A. Sanfeliu, J. Andrade-Cetto, Ubiquitous networking robotics in urban settings, in *Proceedings of IEEE/RSJ IROS Workshop on Network Robots and Systems* (Beijing, Oct 2006), pp. 14–18
16. R. Smith, M. Self, P. Cheeseman, Estimating uncertain spatial relationships in robotics, in *Autonomous Robot Vehicles* (1990), pp. 167–193
17. K.L. Ho, P. Newman, Detecting loop closure with scene sequences. Int. J. Comput. Vis. **74**(3), 261–286 (2007)
18. M. Cummins, P. Newman, FAB-MAP: Probabilistic localization and mapping in the space of appearance. Int. J. Robot. Res. **27**(6), 647–665 (2008)
19. P. Newman, D. Cole, K. Ho, Outdoor SLAM using visual appearance and laser ranging, in *Proceedings of IEEE International Conference on Robotics and Automation* (Orlando, May 2006), pp. 1180–1187
20. K. Fukunaga, *Introduction to Statistical Pattern Recognition*, 2nd edn. (Academic Press, San Diego, 1990)
21. G. Dissanayake, S.B. Williams, H. Durrant-Whyte, T. Bailey, Map management for efficient simultaneous localization and mapping (SLAM). Auton. Robot. **12**(3), 267–286 (2002)

22. T. Vidal-Calleja, A.J. Davison, J. Andrade-Cetto, D.W. Murray, Active control for single camera SLAM, in *Proceedings of IEEE International Conference on Robotics and Automation* (Orlando, May 2006), pp. 1930–1936
23. R. Sim, Stable exploration for bearings-only SLAM. in *Proceedings of IEEE International Conference on Robotics and Automation* (Barcelona, Apr 2005), pp. 2422–2427
24. W. Zhou, J.V. Miro, G. Dissanayake, Information-driven 6D SLAM based on ranging vision, in *Proceedings of IEEE/RSJ IROS Workshop on Network Robots and Systems* (Nice, Sep 2008), pp. 2072–2077
25. H. Kretzschmar, C. Stachniss, G. Grisetti, Efficient information-theoretic graph pruning for graph-based SLAM with laser range finders, in *Proceedings of IEEE/RSJ IROS Workshop on Network Robots and Systems* (San Francisco, Sep 2011), pp. 865–871
26. Y. Ma, S. Soatto, J. Kosecka, S.S. Sastry. *An Invitation to 3-d Vision* (Springer, 2004)
27. E.H. Teniente, J. Andrade-Cetto, FaMSA: Fast multi-scan alignment with partially known correspondences, in *Proceedings of European Conference on Mobile Robotics* (Orebro, Sep 2011), pp. 139–144
28. S. Foix, G. Alenyà, J. Andrade-Cetto, C. Torras, Object modeling using a ToF camera under an uncertainty reduction approach, in *Proceedings of IEEE International Conference on Robotics and Automation* (Anchorage, May 2010), pp. 1306–1312
29. A. Censi. An accurate closed-form estimate of ICP's covariance, in *Proceedings of IEEE International Conference on Robotics and Automation* (Rome, Apr 2007), pp. 3167–3172
30. R. Valencia, E.H. Teniente, E. Trulls, J. Andrade-Cetto, 3D mapping for urban service robots, in *Proceedings of IEEE/RSJ International Conference on Intelligent Robots and Systems* (Saint Louis, Oct 2009), pp. 3076–3081
31. R. Smith, M. Self, P. Cheeseman, A stochastic map for uncertain spatial relationships, in *Proceedings of 4th International Symposium on Robotics Research* (Santa Clara, 1988), pp. 467–474
32. H.F. Durrant-Whyte, Uncertain geometry in robotics. IEEE J. Robot. Autom. **4**(1), 23–31 (1988)
33. R. Chatila, J. Lauomnd, Position referencing and consistent world modeling for mobile robots, in *Proceedings of IEEE International Conference on Robotics and Automation*, vol. 2, (St. Louis, Mar 1985) pp. 138–145
34. J.J. Leonard, H.J.S. Feder, A computationally efficient method for large-scale concurrent mapping and localization, in *Proceedings of 9th International Symposium of Robotics Research*, eds. by J. Hollerbach, D. Koditschek (Salt Lake City, Nov 1999)
35. G. Dissanayake, H. Durrant-Whyte, T. Bailey, A computationally efficient solution to the simultaneous localisation and map building (SLAM) problem. In *Proceedings of IEEE International Conference on Robotics and Automation* (San Francisco, Apr 2000), pp. 1009–1014
36. H. Durrant-Whyte, S. Majumder, M. de Battista, S. Scheding, A Bayesian algorithm for simultaneous localisation and map building. In *Proceedings of 10th International Symposium of Robotics Research* (Lorne, Nov 2001)
37. S. Thrun, Robotic mapping: a survey, in *Exploring Artificial Intelligence in the New Millenium*, eds. by G. Lakemeyer, B. Nebel. World Scientific Series in Robotics and Intelligent Systems (Morgan Kaufmann, 2002)
38. H. Durrant-Whyte, T. Bailey, Simultaneous localisation and mapping (SLAM): Part I the essential algorithms. Robot. Autom. Mag. **13**(2), 99–110 (2006)
39. T. Bailey, H. Durrant-Whyte, Simultaneous localisation and mapping (SLAM): Part II state of the art. Robot. Autom. Mag. **13**(3), 108–117 (2006)
40. T. Bailey, Mobile robot localisation and mapping in extensive outdoor environments. Ph.D. thesis, The University of Sydney, Australian Center for Field Robotics, Sydney, 2002
41. J.D. Tardós, J. Neira, P.M. Newman, J.J. Leonard, Robust mapping and localization in indoor environments using sonar data. Int. J. Robot. Res. **21**(4), 311–330 (2002)
42. J. Andrade-Cetto, T. Vidal-Calleja, A. Sanfeliu, Unscented transformation of vehicle states in SLAM, in *Proceedings of IEEE International Conference on Robotics and Automation* (Barcelona, Apr 2005), pp. 324–329

43. R. Martinez-Cantin, J.A. Castellanos, Unscented SLAM for large-scale outdoor environments, in *Proceedings of IEEE/RSJ International Conference on Intelligent Robots and Systems* (Edmonton, Aug 2005), pp. 328–333
44. S. Thrun, W. Burgard, D. Fox, *Probabilistic Robotics* (MIT Press, Cambridge, 2005)
45. F. Lu, E. Milios, Globally consistent range scan alignment for environment mapping. Auton. Robot. **4**(4), 333–349 (1997)
46. G. Grisetti, C. Stachniss, S. Grzonka, W. Burgard, A tree parameterization for efficiently computing maximum likelihood maps using gradient descent, in *Robotics: Science and Systems III* (Atlanta, Jun 2007), pp. 9:1–9:8
47. G. Grisetti, R. Kummerle, C. Stachniss, W. Burgard, A tutorial on Graph-Based SLAM. Intell. Transp. Syst. Mag. **2**(4), 31–43 (2010)
48. J.B. Folkesson, H.I. Christensen, Graphical SLAM–a self-correcting map, in *Proceedings of IEEE International Conference on Robotics and Automation* (New Orleans, Apr 2004), pp. 383–390
49. J.B. Folkesson, H.I. Christensen, Closing the loop with graphical SLAM. IEEE Trans. Robot. **23**(4), 731–741 (2007)
50. A. Howard, M. Mataric, G. Sukhatme, Relaxation on a mesh: a formalism for generalized localization, in *Proceedings of IEEE/RSJ International Conference on Intelligent Robots and Systems* (Maui, Nov 2001), pp. 1055–1060
51. T. Duckett, S. Marsland, J. Shapiro, Fast, on-line learning of globally consistent maps. Auton. Robot. **12**(3), 287–300 (2002)
52. U. Frese, P. Larsson, T. Duckett, A multilevel relaxation algorithm for simultaneous localization and mapping. IEEE Trans. Robot. **21**(2), 196–207 (2005)
53. E. Olson, J. Leonard, S. Teller, Fast iterative alignment of pose graphs with poor initial estimates, in *Proceedings of IEEE International Conference on Robotics and Automation* (Orlando, May 2006), pp. 2262–2269
54. G. Grisetti, C. Stachniss, W. Burgard, Non-linear constraint network optimization for efficient map learning. IEEE Trans. Intell. Transp. Syst. **10**(3), 428–439 (2009)
55. K. Konolige, G. Grisetti, R. Kuemmerle, W. Burgard, L. Benson, R. Vincent, Psparse pose adjustment for 2d mapping, in *Proceedings of IEEE/RSJ International Conference on Intelligent Robots and Systems* (Taipei, Oct 2010), pp. 22–29
56. R. Kummerle, G. Grisetti, H. Strasdat, K. Konolige, W. Burgard, g2o: a general framework for graph optimization, in *Proceedings of IEEE International Conference on Robotics and Automation* (Shanghai, May 2011), pp. 3607–3613
57. M. Kaess, H. Johannsson, R. Roberts, V. Ila, J. Leonard, F. Dellaert, iSAM2: incremental smoothing and mapping using the bayes tree. Int. J. Robot. Res. **31**(2), 216–235 (2012)
58. G. Klein, D. Murray, Parallel tracking and mapping for small AR workspaces, in *Proceedings of IEEE International Symposium on Mixed and Augmented Reality* (Nara, Nov 2007), pp. 225–234
59. E. Eade, T. Drummond, Monocular SLAM as a graph of coalesced observations, in *Proceedings of IEEE International Conference on Computer Vision* (Rio de Janeiro, Oct 2007), pp. 1–8
60. P. Piniés, L.M. Paz, D. Galvez-Lopez, J.D. Tardós, CI-Graph simultaneous localization and mapping for three-dimensional reconstruction of large and complex environments using a multicamera system. J. Field Robot. **27**(5), 561–586 (2010)
61. N. Kai, D. Steedly, F. Dellaert, Tectonic SAM: exact, out-of-core, submap-based SLAM, in *Proceedings of IEEE International Conference on Robotics and Automation* (Barcelona, Apr 2005), pp. 1678–1685
62. M. Bosse, P. Newman, J. Leonard, S. Teller, Simultaneous localization and map building in large-scale cyclic environments using the atlas framework. Int. J. Robot. Res. **23**(12), 1113–1139 (2004)
63. G. Sibley, C. Mei, I. Reid, P. Newman, Vast-scale outdoor navigation using adaptive relative bundle adjustment. Int. J. Robot. Res. **29**(8), 958–980 (2010)
64. J. Neira, J.D. Tardós, Data association in stochastic mapping using the joint compatibility test. IEEE Trans. Robot. Autom. **17**(6), 890–897 (2001)

65. J.A. Castellanos, J.M. Martínez, J. Neira, J.D. Tardós, Experiments in multisensor mobile robot localization and map building, in *3rd IFAC Symposium on Intelligent Autonomous Vehicles* (Madrid, 1998), pp. 173–178
66. J. Andrade-Cetto, A. Sanfeliu, Temporal landmark validation in CML, in *Proceedings of IEEE International Conference on Robotics and Automation* (Taipei, Sep 2003), pp. 1576–1581
67. N. Sunderhauf, P. Protzel, Switchable constraints for Robust Pose Graph SLAM, in *Proceedings of IEEE/RSJ International Conference on Intelligent Robots and Systems* (Algarve, Oct 2012), pp. 1879–1884
68. N. Sunderhauf, P. Protzel, Towards a robust back-end for pose graph SLAM, in *Proceedings of IEEE International Conference on Robotics and Automation* (Saint Paul, Minessota, May 2012), pp. 1254–1261
69. P. Agarwal, G. Tipaldi, L. Spinello, C. Stachniss, W. Burgard, Robust map optimization using dynamic covariance scaling, in *Proceedings of IEEE International Conference on Robotics and Automation* (Karlsruhe, May 2013), pp. 62–69
70. E. Olson, P. Agarwal, Inference on networks of mixtures for robust robot mapping. Int. J. Robot. Res. **32**(7), 826–840 (2013)
71. Y. Latif, C. Cadena, J. Neira, Robust loop closing over time for pose graph SLAM. Int. J. Robot. Res. **32**(14), 1611–1626 (2013)
72. L. Carlone, A. Censi, F. Dellaert, Selecting good measurements via L1 relaxation: a convex approach for robust estimation over graphs, in *Proceedings of IEEE/RSJ International Conference on Intelligent Robots and Systems* (Chicago, Sep 2014), pp. 2667–2674
73. Y. Latif, C. Cadena, J. Neira, Robust graph SLAM back-ends: a comparative analysis, in *Proceedings of IEEE/RSJ International Conference on Intelligent Robots and Systems* (Chicago, Sep 2014), pp. 2683–2690
74. N. Sunderhauf, P. Protzel, Switchable Constraints vs. Max-Mixture models vs. RRR-A comparison of three approaches to robust pose graph SLAM, in *Proceedings of IEEE International Conference on Robotics and Automation* (Karlsruhe, May 2013), pp. 5198–5203
75. T. Vidal-Calleja, Visual navigation in unknown environments. Ph.D. thesis, UPC, Barcelona, Jul 2007
76. A. Nüchter, K. Lingemann, J. Hertzberg, H. Surmann, 6D SLAM-3D mapping outdoor environments. J. Field Robot. **24**(8–9), 699–722 (2007)
77. P.J. Besl, N.D. McKay, A method for registration of 3D shapes. IEEE Trans. Pattern Anal. Mach. Intell. **14**(2), 239–256 (1992)
78. D.M. Cole, P.M. Newman, 3D SLAM in outdoor environments, in *Proceedings of IEEE International Conference on Robotics and Automation* (Orlando, May 2006), pp. 1556–1563
79. C. Leung, S. Huang, N. Kwok, G. Dissanayake, Planning under uncertainty using model predictive control for information gathering. Robot. Auton. Syst. **54**(11), 898–910 (2006)
80. T. Vidal-Calleja, A. Sanfeliu, J. Andrade-Cetto, Action selection for single camera SLAM. IEEE Trans. Syst. Man, Cybern. B **40**(6), 1567–1581, (2010). (Dec 2010)

Chapter 4
Path Planning in Belief Space with Pose SLAM

The probabilistic belief networks that result from standard feature-based simultaneous localization and map building methods cannot be directly used to plan trajectories. The reason is that these methods produce a sparse graph of landmark estimates and their probabilistic relations, which is of little value to find collision free paths for navigation. These graphs can be enriched with obstacle or traversability related information, but at the expense of an increase in complexity. Moreover, the resulting planning methods typically do not exploit the sources of uncertainty encoded in the maps. In contrast, we argue in this Chapter that Pose SLAM graphs can be directly used as belief roadmaps [1, 2] and, thus, used for path planning under uncertainty. The method we present in this Chapter devises optimal navigation strategies by searching for the path in the pose graph with the lowest accumulated robot pose uncertainty, i.e., the most reliable path to the goal.

Aside from applications such as the reconstruction of archaeological sites [3] or the inspection of dangerous areas [4], the final objective for an autonomous robot is not to build a map of the environment, but to use this map for navigation. Whereas in the seminal approaches to SLAM [5] only few tens of landmarks could be managed, state of the art approaches can now efficiently manage thousands of landmarks [6–8] and build maps over several kilometers [9]. However, for efficiency reasons, most SLAM algorithms represent the environment using a sparse set of features. Unfortunately, this representation cannot be directly used for collision-free path planning since it does not provide much information about which routes in the map have been previously traversed safely, or about the nature of the obstacles it represents. Those sparse models could be somehow enriched with obstacle or traversability related information [10–12], but at the expense of an increase in complexity. For instance, the traversability map shown in Chap. 3 requires to further process the volumetric map obtained with Pose SLAM.

The problem of finding paths to reach distant locations is addressed in the motion planning literature. However, the research in motion planning typically assumes

R. Valencia and J. Andrade-Cetto, *Mapping, Planning and Exploration with Pose SLAM*, Springer Tracts in Advanced Robotics 119,
DOI 10.1007/978-3-319-60603-3_4

deterministic setups where a perfect model of the environment is available and where the configuration of the robot is perfectly known too. The most successful methods are based on randomized sampling [13, 14], in which collision-free configurations are stochastically drawn and where, if possible, neighbor samples are connected forming a roadmap. This roadmap is later used to find a path between any two given configurations. Some approaches have addressed the problem of optimizing the quality of this path, mainly focusing on reducing the path length [15, 16].

Some extensions have been introduced to deal with uncertainties in the model of the environment [17], in the robot configuration [18], in the effect of robot actions [19], or in the effect of actions and measurements [20]. The extension that best matches the stochastic nature of SLAM is the Belief Roadmap (BRM) [1, 2]. In this approach, the edges defining the roadmap include information about the uncertainty change when traversing such an edge. However, the main drawback of the BRM is that it still assumes a known model of the environment, which is in general not available in real applications. In this work, we aim to overcome the limitation of BRMs noting that the map generated by Pose SLAM [21], or any other delayed-state SLAM method [22–24], can be directly used as a belief roadmap (see Fig. 4.1).

In a semi-autonomous scenario where a human initially drives the robot through a set of interest way points, the outcome of Pose SLAM is a graph of obstacle-free paths in the area where the robot has been operated. Using this graph for navigation allows to have an infrastructure-free automated guided vehicle, as those widely used for surveillance, for material distribution in factories, or for drug delivery in hospitals [25, 26]. An added advantage is that Pose SLAM is agnostic with respect to the sensor modalities used, which facilitates its application in different environments and robots, and the paths stored in the map satisfy constraints not easy to model in the robot controller, such as the existence of restricted regions, or the right of way along paths. Deviations from these paths might result in an inconvenience for other operations carried out in the factory or hospital. Thus, a robot that can adequately choose the correct path from a set of previously traversed ones, or their combination, is desirable for such applications. However, in those scenarios, the use of a robot is only practical if it is able to navigate without becoming lost, i.e., without requiring the intervention of the operator. The research reported in this Chapter addresses this issue, providing the safest path from one waypoint to another.

In most cases, any two poses are connected by different paths, and navigating through one or the other would entail different probabilities of becoming lost. A path through areas where sensory data is not reliable means higher risk of deviating from the path to follow during execution (see Fig. 4.2). In this Chapter, we show that, using Pose SLAM, we can plan in the belief space to obtain paths that take into account the uncertainties along the path. The key idea behind our method is that, in Pose SLAM, highly informative areas of the environment result in poses in the graph with low uncertainty. Thus, paths through safer areas in the sense of being reliably localized can be selected considering only the pre-computed uncertainties encoded in the graph.

From the point of view of SLAM, this method constitutes a step forward to actually use the output of the mapping process for path planning. From the point of view of

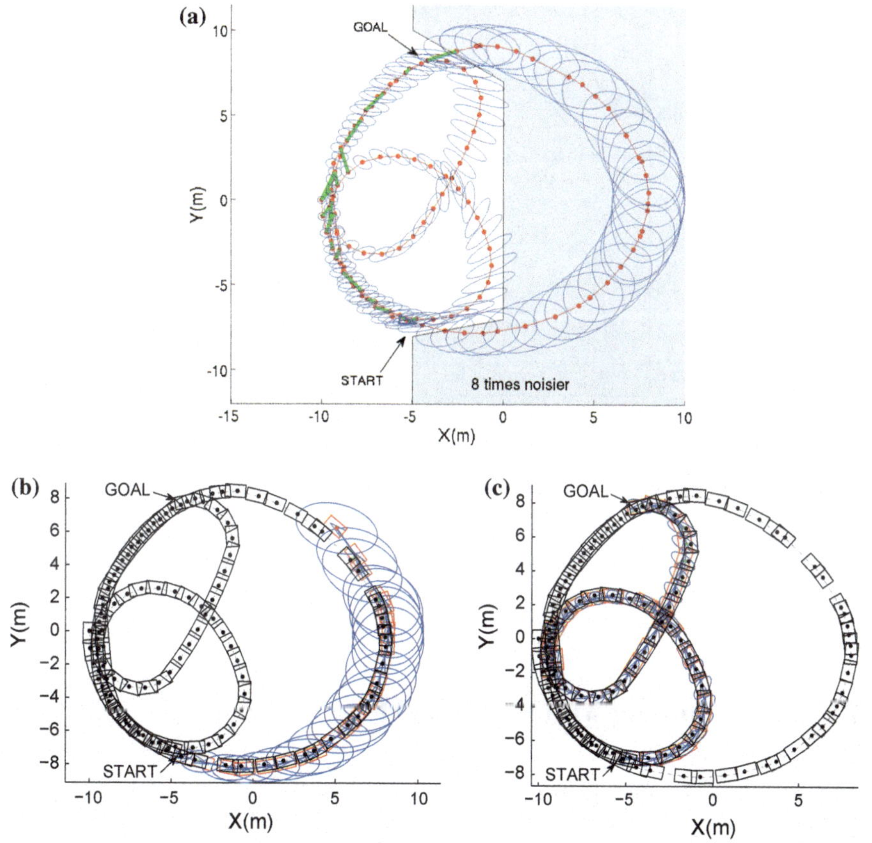

Fig. 4.1 Path planning using the map generated by the Pose SLAM algorithm. **a** The Pose SLAM map. The *red dots and lines* represent the estimated trajectory, and the *green lines* indicate loop closure constraints established by registering sensor readings at different poses. **b** A plan in configuration space would produce the shortest path to the goal. At one point during path execution, sensor registration fails and the robot gets lost. This happens when the robot is outside the sensor registration area for a given waypoint in the tracked trajectory. The areas around each pose where registration is possible are represented by *rectangles*. **c** A plan in belief space produces the minimum uncertainty path to the goal. Plans with low uncertainty have higher probability of success

motion planning, this approach contributes with a method to generate belief roadmaps without resorting to stochastic sampling on a pre-defined environment model.

The rest of the Chapter details thc proposed approach to perform path planning. As we mentioned, this method reinterprets the Pose SLAM map as a set of samples in belief space and, in Sect. 4.1 we describe how to plan using a roadmap defined on these samples. Next, Sect. 4.2 presents the planning algorithm. In Sect. 4.3, this new planning approach is tested with datasets and a real world experiment. Finally, Sect. 4.4 shows the related work to our approach.

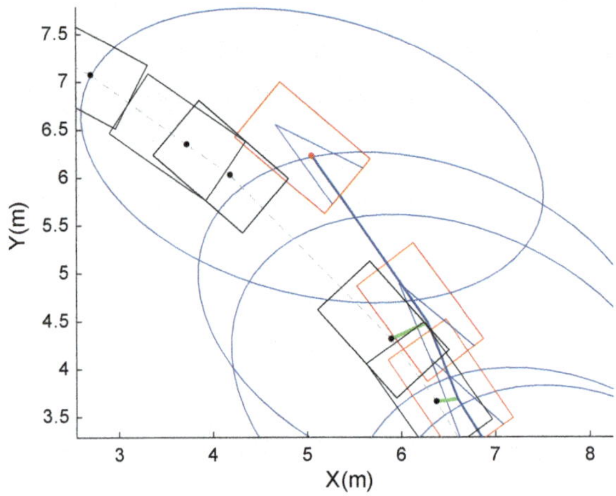

Fig. 4.2 Zoomed view of a region along the shortest path in Fig. 4.1 where the robot gets lost. Bad localization on this path leads the robot to deviate from the next waypoint, producing failed sensor registration. The *rectangles* indicate the areas where sensor registration is reliable (shown in *black* for the poses in the map and in *red* for the poses in the executed trajectory), the *green lines* represent sensor registration links between poses in the executed trajectory and those in the map, and the *blue lines* and ellipses represent the localization estimates for the executed trajectory

4.1 Path Planning with Pose SLAM

We are in the quest for a path $p = r_{1:T}$ that would drive the robot from its current configuration $s_t = r_1$ to a goal configuration $g = r_T$, that, for the applications considered in this work, is always included in the Pose SLAM graph. In this quest, we assume that the robot is equipped with a local planner able to drive the robot to nearby configurations.

Moreover, we assume maximum likelihood actions and measurements, as it is usual when planning in belief space [27]. With this, the mean estimate after a sequence of controls will lie at the mean of a node in the graph and the observation previously obtained at that position will be repeated. In consequence, the proposed planning approach only needs to consider poses in the graph, which are guaranteed to be collision free.

Given that candidate paths lie on the top of this graph, after path execution the final robot uncertainty will be close to the original marginal at that node. Thus, a cost function that only evaluates the belief state at the goal is unsuitable and we are interested instead in determining reliable paths, i.e., paths where the robot has low probability of becoming lost. Herein, we assume that the probability of becoming lost is directly related with the increments in the uncertainty in the robot positioning since uncertainty decrements can only result in a better track of the path.

We propose to achieve this objective in four stages. First, we increase the connectivity of the Pose SLAM graph so that paths combining different exploration sequences can be considered. Next, we propose a principled way to evaluate the uncertainty of transitions between nodes. Then, we use this uncertainty measure to define the cost of a path as its mechanical work in the uncertainty surface. Finally, we integrate the three previous points to derive the proposed path planning algorithm.

4.1.1 Increasing Graph Connectivity

The graph used for path planning is initialized with the odometry edges of the Pose SLAM graph. However, we allow the local planner to attempt connections to other neighboring poses. In this way, the planner can switch among different exploration sequences in the quest for an optimal path.

Extra links are included for neighboring nodes with high probability of being close to each other and, thus, likely to be reachable using the local planner. This is done with the technique shown in Sect. 3.1.3 from Chap. 3.

To determine such poses, we estimate the relative displacement, d, from any robot pose x_k to any other pose x_i as a Gaussian with parameters given by Eqs. 3.16 and 3.17. Then, marginalizing the distribution of the displacement, d, along each one of its dimensions, t, we get a one-dimensional Gaussian distribution $\mathcal{N}(\mu_t, \sigma_t^2)$ that allows to compute the probability of pose x_i being closer than v_t to pose x_k along such dimension with Eq. 3.18. If for all dimensions, p_t is above a given threshold s, then configuration x_i is considered close enough to configuration, x_k.

Pose SLAM [21] computes the set of neighbors for each pose during mapping in logarithmic time, organizing the poses in a tree. For other delayed-state systems, though, the computation of the set of neighbors require to compute the marginal covariances and the cross correlations between all pairs of poses, which is computationally expensive.

Observe that the proposed approach remains agnostic about the properties of the local planner and, thus, neighboring poses are searched in a rectangle around the current pose in configuration space. If we had information about the kinematic constraints of the robot, or even the distribution of obstacles around the robot, we could reduce the search for neighboring poses to smaller areas. For instance, for a car-like robot, we could focus the search for neighbors into triangular areas in front of and behind the robot since these are the regions including the kinematically-feasible neighboring poses. Also, for a robot that can only safely move forward due to the arrangement of its obstacle detection sensors, only neighboring poses in front of the robot need to be detected. In any case, the size of the area where to look for neighbors is limited by the accuracy of the local planner, which typically relies in odometry readings.

Independently of the location and size of the area used to search for neighbors, an edge is added to the path planning graph only if the kinematic constraints enforced by the local planner allow to reach the nearby node. Despite this verification, the local path might still result unfeasible during path execution, mainly due to the presence of obstacles. Then, the problematic edge can be removed from the graph and a re-planning process can be triggered.

4.1.2 Uncertainty of a Path Step

To track a path within the pose SLAM map, the robot departs from an initial pose estimate and predicts a new estimate taking into account the relative motion constraint between two consecutive nodes in the path. Such a tracker would update this new estimate registering current sensor readings to those stored in the corresponding map node. Thus, every pair of poses in a planned path, say r_{k-1} and r_k, would match two poses in the Pose SLAM graph, say x_i and x_j. The command u_k that drives the robot from r_{k-1} to r_k and the associated motion noise $\mathbf{\Sigma}_u$ are provided by the local planner. Since, during the planning process the actual sensor readings are not available, the update of this estimate is made assuming that at the end of the step from r_{k-1} to r_k, r_k and x_j will be coincident and, thus, $z_{kj} = r_k - x_j \sim \mathcal{N}(\mathbf{0}, \mathbf{\Sigma}_{jj})$ where $\mathbf{\Sigma}_{jj}$ is the marginal covariance of the pose x_j.

To evaluate the change in uncertainty introduced in the motion from r_{k-1} to r_k, we must look at the conditional distribution $p(r_k|r_{k-1}, u_k, z_{kj})$ that is obtained from the estimation of the joint state (r_{k-1}, r_k). This estimation can be obtained using, for instance, an EKF [28], as we show next.

The cost of traversing a link from node r_{k-1} to node r_k is proportional to the conditional entropy at node j given full confidence about node i, $H(r_k|r_{k-1})$, which for Gaussians is proportional to

$$H(r_k|r_{k-1}) \propto |\bar{\mathbf{\Sigma}}_{k,k} - \bar{\mathbf{\Sigma}}_{k,k-1}\bar{\mathbf{\Sigma}}^{-1}_{k-1,k-1}\bar{\mathbf{\Sigma}}_{k-1,k}|, \tag{4.1}$$

where the marginals and cross-correlations are extracted from $\bar{\mathbf{\Sigma}}$, the covariance of the compound localization estimate (r_{k-1}, r_k). Equation 4.1 is a measure of the robot's ability to safely track its position during path execution. To compute both marginals and cross correlation terms in Eq. 4.1 we need to track localization estimates of the previous and current robot poses x_i and x_j. To this end, the compound localization estimate (x_i, x_j) can be computed with the Extended Kalman Filter (EKF), with the particularity that every EKF update is given by sensor registration with the Pose SLAM graph at node j, taking into account its marginal covariance Σ_{jj}.

Nevertheless, going a step further, this uncertainty can be more elegantly evaluated using an EIF instead. With an EIF, (r_{k-1}, r_k) is estimated as $\mathcal{N}^{-1}(\bar{\eta}, \bar{\mathbf{\Lambda}})$ and in the prediction step the joint information matrix is updated as [23]

$$\begin{bmatrix} \bar{\Lambda}_{k-1k-1} + \mathbf{F}\,\mathbf{\Sigma}_u^{-1}\,\mathbf{F}^{\top} & -\mathbf{F}^{\top}\mathbf{\Sigma}_u^{-1} \\ -\mathbf{\Sigma}_u^{-1}\,\mathbf{F} & \mathbf{\Sigma}_u^{-1} \end{bmatrix}, \tag{4.2}$$

with $\mathbf{F}$ the Jacobian of f with respect to r_{k-1} evaluated at the mean, and where $\bar{\Lambda}_{k-1k-1}$ is obtained marginalizing from the previous $\bar{\mathbf{\Lambda}}$. In the correction step the measurement z_{kj} is used to update the information matrix as

$$\bar{\mathbf{\Lambda}} = \begin{bmatrix} \bar{\Lambda}_{k-1k-1} + \mathbf{F}\,\mathbf{\Sigma}_u^{-1}\,\mathbf{F}^\top & -\mathbf{F}^\top\mathbf{\Sigma}_u^{-1} \\ -\mathbf{\Sigma}_u^{-1}\,\mathbf{F} & \mathbf{\Sigma}_u^{-1} + \mathbf{\Sigma}_{jj}^{-1} \end{bmatrix} . \tag{4.3}$$

With this, the uncertainty of r_k given full confidence about r_{k-1} can be evaluated as

$$H(r_k|r_{k-1}) \propto \frac{1}{|\bar{\Lambda}_{k|k-1}|} , \tag{4.4}$$

where $\bar{\Lambda}_{k|k-1}$ is the information matrix for the conditional $p(r_k|r_{k-1}, u_k, z_{kj})$ obtained from $\bar{\Lambda}$.

Fortunately enough, conditioning in information form is dual to marginalization in covariance form, which saves us from computing the state covariance of the tracker and the Schur complement to condition the covariance, as done in Eq. 4.1. Thus, using the EIF estimation, Eq. (4.4) simply evaluates to

$$U_k = \frac{1}{|\mathbf{\Sigma}_u^{-1} + \mathbf{\Sigma}_{jj}^{-1}|} . \tag{4.5}$$

We can safely assume $\mathbf{\Sigma}_u$ to be non-degenerate and, thus, the determinant in Eq. (4.5) would never be null and U_k will be always well-defined. Note that we use a measure of uncertainty change derived from the determinant of the covariance matrix which is related to the entropy of $p(r_k|r_{k-1}, u_k, z_{kj})$, and ultimately, to the uncertainty hyperellipsoid defined by this matrix. A trace-based uncertainty measure [1] can be used as well, without affecting the overall planning algorithm.

Note that this measure of uncertainty change is computed independently of the estimation of the robot pose at that step and, thus, this formulation saves us from actually implementing the EIF to track the path. This does not imply that the robot must have an identical belief to a pre-existing node in the optimized Pose SLAM graph. What it implies is that the computation of the information gain is independent of the current belief, as long as maximum likelihood actions and measurements are considered. This is especially relevant to allow planning with different initial beliefs. Moreover, as long as the graph does not change, the uncertainty measure for all transitions can be precomputed from the Pose SLAM graph and re-used to plan different paths. This is similar to what is done in [1], factorizing the covariance update, but simplified thanks to the use of the information form.

4.1.3 Minimum Uncertainty Along a Path

Next, we propose a cost function that considers cumulative relative uncertainty during localization. In principle, only increments in uncertainty are problematic for the quality of the localization since any decrement in uncertainty can only result in a better estimate. Therefore, finding paths that accumulate the least uncertainty can be seen as searching for a path of minimal mechanical work [29] in an uncertainty surface [30] over the space of robot poses, where the uncertainty of a path step is computed using the criterion described in Sect. 4.1.2.

Given a discrete path $p = r_{1:T}$, we define its mechanical work in the uncertainty surface as the sum of positive increments of individual step costs

$$W(r_{1:T}) = \sum_{k=2}^{T} \Delta U_k^+, \tag{4.6}$$

with

$$\Delta U_k^+ = \begin{cases} \Delta U_k & \Delta U_k > 0, \\ 0 & \Delta U_k \leq 0, \end{cases} \tag{4.7}$$

and

$$\Delta U_k = U_k - U_{k-1} \tag{4.8}$$

and where, by convention, $U_1 = 0$, to include the uncertainty of the first step of the path in W. Note that, the initial uncertainty of the robot is not included in W since it would result in a constant offset for the cost of all alternative paths. Moreover, since the costs are non-negative, there is always an acyclic minimum cost path to each reachable node in the map.

This strategy prefers short paths with possibly steep uncertainty changes over much longer paths with gentle oscillations of uncertainty, thus avoiding the accumulation of small chances of becoming lost over large trajectories. In this way, the proposed strategy is a mechanism that adequately balances path length with uncertainty change.

4.2 The Pose SLAM Path Planning Algorithm

To find the optimal path, Dijkstra's algorithm is implemented with the cost function defined in Sect. 4.1.3. It is formally described in Algorithm 1. It implements a minimum uncertainty path search among paths tracked on the poses in a Pose SLAM graph. The algorithm takes as inputs the Pose SLAM graph M and the goal pose, g, which is assumed in M. Should this not be the case, the closest pose in the graph to g (in configuration space) is used as a goal. We first initialize a set Q with all

Algorithm 1: Path planning with Pose SLAM.

PoseSLAMPathPlanning(M,g)

input : M: The graph computed by Pose SLAM.
g: The goal pose.
output: p: Minimum uncertainty path to g.

```
1  Q ← POSES(M)
2  forall the n ∈ Q do
3      W[n] ← ∞
4      V[n] ← 0
5  s_t ← CURRENTPOSE(M)
6  W[s_t] ← 0
7  U[s_t] ← 0
8  repeat
9      i ← EXTRACTMIN(Q, W)
10     if i ≠ g and W[i] ≠ ∞ then
11         N ← NEIGHBORS(M, i)
12         forall the j ∈ N do
13             (u, Σ_u) ← LOCALPLANNER(x_i, x_j)
14             if u ≠ ∅ then
15                 Σ_jj ← MARGINALCOVARIANCE(M, j)
16                 U = 1/|Σ_u^{-1} + Σ_jj^{-1}|
17                 ΔU = U − U[i]
18                 if ΔU > 0 then
19                     W' = W[i] + ΔU
20                 else
21                     W' = W[i]
22                 if W' < W[j] then
23                     W[j] ← W'
24                     V[j] ← i
25                     U[j] ← U
26 until i = g or W[i] = ∞ or Q = ∅;
27 p ← ∅
28 if i = g then
29     c ← g
30     while c ≠ 0 do
31         p ← {c} ∪ p
32         c ← V[c]
33 RETURN p
```

the nodes in the graph (Line 1) and establish an initial cost W for the path to each node (Line 3) and a fake predecessor V for each node (Line 4). Then, the cost to reach the starting configuration is set to 0 (Lines 5 to 7). At this point the algorithm enters in a loop until the goal is reached or the reachable region from the start configuration is fully explored (Lines 8 to 26). At each iteration of the loop, we extract the node i with minimum cost from Q (Line 9). If this is not the goal (Line 10), we perform breadth first search on the neighbor nodes to i (Line 11). The neighboring nodes are determined using the procedure given in Sect. 4.1.1 that takes into account the uncertainty in the pose estimates. For each one of the possible transitions to neighbors, we use the local planner to determine if the transition is possible and to compute the expected motion uncertainty (Line 13). Using this uncertainty and the marginal covariance for the target pose (Line 15) we compute the step uncertainty as described in Sect. 4.1.2 (Line 16). Then, Line 17 computes the uncertainty increment for a motion from node i to node j. If this increment is positive, it is added to the path cost (Line 19). Otherwise, this step does not contribute to the overall path cost. If the new path to j is lower than the best known until that moment, the cost to reach j is updated (Line 23), we set i as the predecessor of j (Line 24), and we store the cost for the path to the node (Line 25). In the case of paths with equal cost, shorter ones are preferred and, since the costs of individual steps are non-negative, the considered paths never include cycles. If the goal is reached, the minimum uncertainty path to the goal is reconstructed using the chains to predecessor nodes stored in V (Lines 30 to 32). If the goal is determined to be non-reachable from the start configuration, an empty path is returned.

Without considering the cost of recovering the marginal covariances, the asymptotic cost of the algorithm is $O(e \log^2 n)$ with e the number of edges in the graph (i.e., the number of neighboring pose pairs) and n the number of nodes in the graph. This cost assumes that the nodes in Q are organized into a heap where the extraction of the minimum element is constant time and the update of the cost of an element is logarithmic. Moreover, it also assumes that poses are organized into a tree so that neighboring poses can be determined logarithmically [21]. If this search is performed linearly the cost increases to $O(e\, n \log n)$.

Note that, when planning we do not need to maintain a localization estimate, but still we need to simulate registration with the map, for which the diagonal blocks of the covariance matrix are needed (Line 16). When using the Pose SLAM algorithm [21], these diagonal blocks are directly available [31], but this is not the case in other approaches [22–24]. In these cases, the most efficient way to compute the marginals is to invert the whole information matrix before starting to plan. One can efficiently invert it taking advantage of its sparsity using, for instance, sparse supernodal Cholesky decomposition [32]. For large-scale problems, however, this strategy becomes prohibitively expensive and we have to resort to approximations of the marginal covariances obtained using, for instance, Markov blankets [33].

Finally, should the map change significantly during path execution (i.e., a new highly informative loop closure is found), the Pose SLAM algorithm performs a full state update and re-planning is enforced.

4.3 Experimental Results

In order to evaluate the planning strategy presented in this Chapter we show results with four data sets and with a real robot navigation experiment. The results with the data sets were obtained with a Matlab implementation running on an Intel Core2 Quad system at 3 GHz with 4 GB of memory. For the real robot navigation experiment, the system was implemented using the Robot Operating System (ROS) [34] on our 4-wheel robot Teo, a Segway RMP400 platform.

4.3.1 Synthetic Dataset

In the first experiment, we simulate a robot moving over a given trajectory with several loops. In the simulation, the motion of the robot is measured with an odometric sensor whose error is 5% of the displacement in x and y, and 0.0175 rad in orientation. A second sensor is simulated to establish links between any two poses closer than ± 1.25 m in x, ± 0.75 m in y, and ± 0.26 rad in orientation, with noise covariance $\mathbf{\Sigma}_z = \text{diag}(0.2\,\text{m}, 0.2\,\text{m}, 0.009\,\text{rad})^2$. The initial uncertainty in the robot pose is set to $\mathbf{\Sigma}_0 = \text{diag}(0.1\,\text{m}, 0.1\,\text{m}, 0.09\,\text{rad})^2$. Nearby poses are detected with a very permissive $s = 0.1$. With such value of s we avoid missing any neighbor. Finally, the minimum information gain, γ is set to 3 nats.

Figure 4.1a shows the final map as estimated by the Pose SLAM algorithm. The shadowed area indicates harsher navigation conditions with odometry and loop closure errors increased by a factor of 8. This noisier area simulates a part of the environment where constraints between poses are harder to establish.

After building the map using Pose SLAM we planned a path to a particular goal selected from the nodes in the map. Figure 4.1b shows the trajectory to the goal using a shortest path criterion, and Fig. 4.1c shows the trajectory obtained when using the

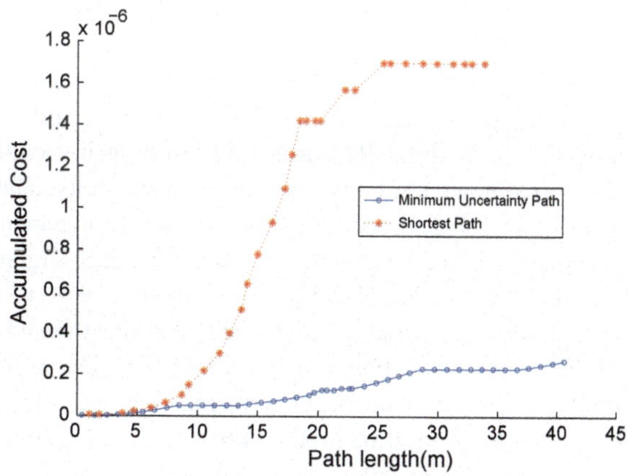

Fig. 4.3 Accumulated cost along the shortest (*red*) and minimum uncertainty (*blue*) paths in the simulated experiment

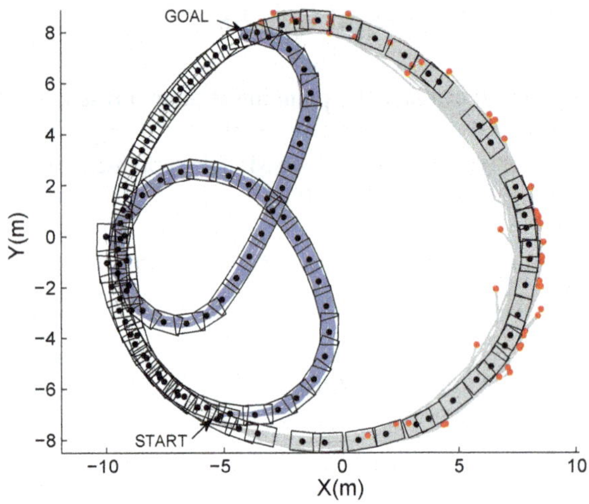

Fig. 4.4 Monte Carlo realization of the simulated experiment. The minimum uncertainty path guarantees path completion during localization as indicated by the *blue* trajectories. The *red dots* indicate on the other hand, the points where the robot gets lost due to a missed sensor registration, while executing the shortest path

minimum uncertainty criterion introduced in Sect. 4.1.3, which avoids the noisier area in the environment.

Figure 4.3 shows a plot of the accumulated cost along the two trajectories. The accumulated uncertainty of the shortest path is significantly larger than that of the minimum uncertainty path. Therefore, following this second trajectory there is increased guarantee that the robot will be better localized all along the path and will less likely get into trouble, for instance, of getting lost. This is verified in Fig. 4.4 that shows a Monte Carlo realization of the this experiment with 100 runs. Navigation through the shortest path reached the goal only 45% of the times due to failed sensor registration along the path, whereas navigating over the minimum uncertainty path always reached the final destination since the trajectory avoids the noisier area in the environment.

4.3.2 *Indoor Dataset*

To test the performance of the algorithm on an indoor data set we choose the data collected at the Intel Research Lab building in Seattle [35]. The dataset includes 26915 odometry readings and 13631 laser scans. The laser readings were used to generate scan-based odometry and to assert loop closures, by aligning them using an incremental closest point (ICP) scan matching algorithm [22]. In this case, only links between poses closer than ± 1 m in x and y, and ± 0.35 rad in orientation were considered reliable. These are also the thresholds used to determine neighboring poses when planning with $s = 0.1$. The robot odometry and the relative motion computed from laser scan matches were modeled with noise covariances $\mathbf{\Sigma}_u = \text{diag}(0.05\,\text{m}, 0.05\,\text{m}, 0.03\,\text{rad})^2$ and $\mathbf{\Sigma}_z = \text{diag}(0.05\,\text{m}, 0.05\,\text{m}, 0.009\,\text{rad})^2$,

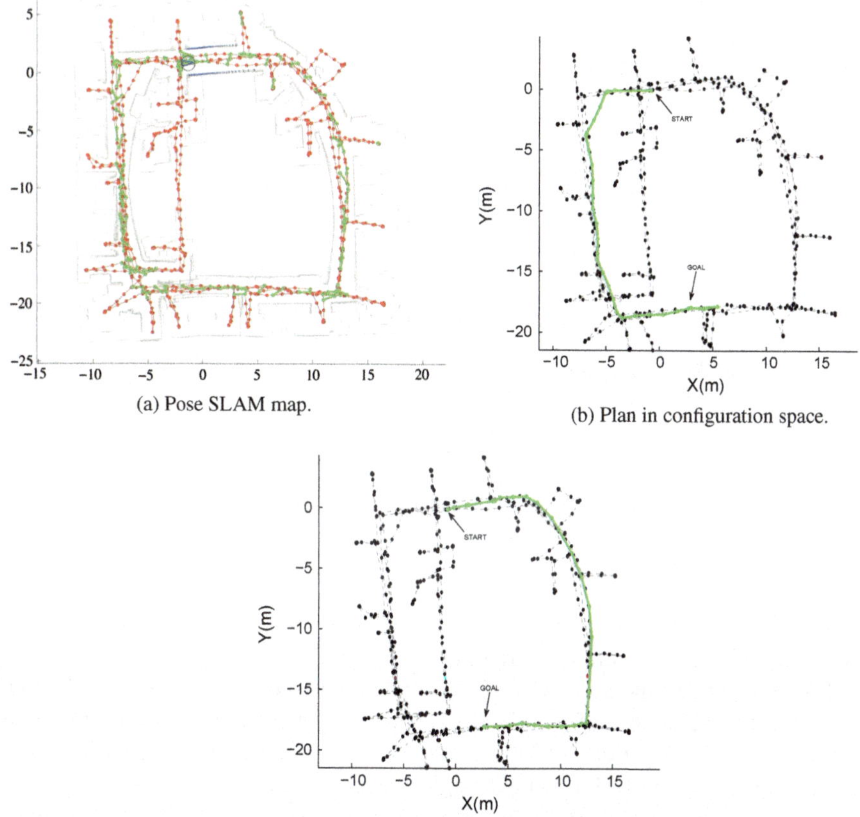

(a) Pose SLAM map.

(b) Plan in configuration space.

(c) Plan in belief space.

Fig. 4.5 Path planning over the Intel dataset. **a** Pose SLAM map built with encoder odometry and laser scans. The *blue arrow* indicates the final pose of the robot and the *black* ellipse the associated covariance at a 95% confidence level. **b** Planning in configuration space we obtain the shortest path to the goal on the underlying Pose SLAM graph. **c** Planning in belief space we obtain the minimum uncertainty path to the goal

respectively and the minimum information gain was $\gamma = 4.5$ nats. Figure 4.5a shows the path estimated by Pose SLAM together with the laser scans associated to each of the stored poses in light gray.

This map is the departing point of the planning algorithm and the goal is to connect two poses on opposite sides of the building. Frames (b) and (c) in Fig. 4.5 show the shortest and minimum-uncertainty paths between the two poses. The apparent overshoot of the shortest path to the goal is due to the fact that the robot has to execute a 180 deg turn at the end of the path to align with the goal since sudden changes in orientations are not allowed by the kinematic constraints assumed for the robot. This

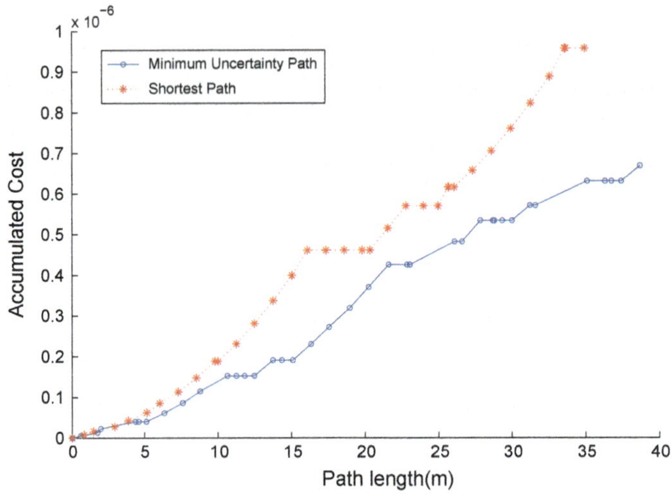

Fig. 4.6 Accumulated cost versus the path length for the shortest path (*red*) and minimum uncertainty path (*blue*) in the Intel experiment

rotation is only possible few meters away of the goal, in front of a door where many poses with the robot at different orientations accumulate.

Figure 4.6 shows the accumulated cost along the two paths. We can note that the accumulated uncertainty of the shortest path is larger than that for the minimum uncertainty path. Therefore, following this second path the robot is better localized all the time at the cost of following a slightly larger path.

To test the efficiency of the method, Fig. 4.7 shows the execution time and memory footprint for planning as a function of problem size, varying the number of poses in the Intel map. Since the most expensive step of the algorithm is the recovery of the marginal covariances, we applied two different strategies to recover them: recovering the whole $\mathbf{\Sigma}$ and recovering it column-wise as needed during planning. The continuous lines in Fig. 4.7a show the execution time needed to recover the marginals as a function of problem size, whereas the dashed lines show the execution time of the whole planning algorithm also as a function of problem size. The figure shows that recovering the whole matrix is computationally more efficient at the expense of increased memory space. On the contrary, on-the-fly computation of matrix columns results in repeated computations slowing down planner performance. The execution cost of re-planning when a graph edge is found to be non-traversable is reduced to the difference between the continuous and the dashed lines in Fig. 4.7a since the map does not change and, thus, the marginal covariances do not need to be re-computed.

Fig. 4.7 Plots of execution time and memory footprint when planning with different subsets of the Intel map and employing two different strategies to recover marginals. **a** Execution time needed to recover only the marginals (*continuous line*) and for the whole planning algorithm (*dashed line*). **b** Memory footprint for marginal recovery

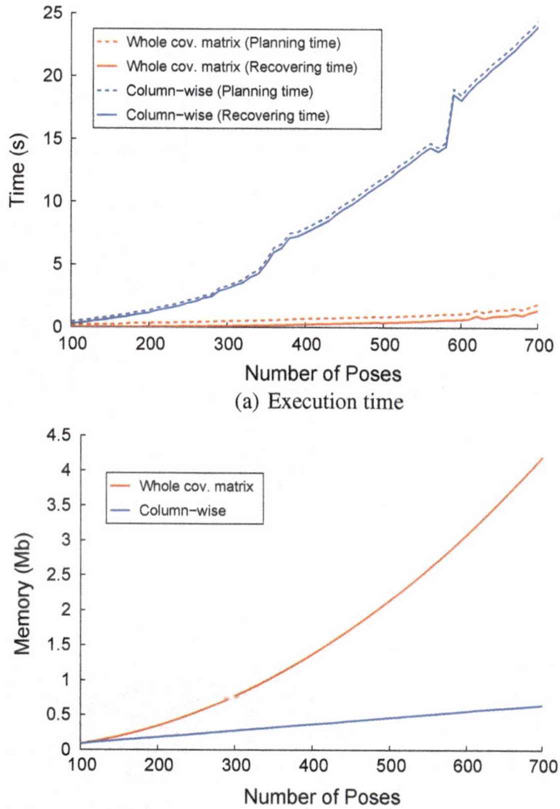

(a) Execution time

(b) Memory footprint

4.3.3 Large Scale Dataset

To demonstrate scalability, we tested our approach with a much larger map, for which memory space is a constraint. To this end, we planned paths using the simulated Manhattan data set [36] that includes over 10000 poses. In this experiment, noise covariances for robot odometry and the relative-pose measurements were set to $\mathbf{\Sigma}_u = \mathbf{\Sigma}_z = \text{diag}(0.05\,\text{m}, 0.05\,\text{m}, 0.03\,\text{rad})^2$, the threshold to detect neighboring poses was $s = 0.1$ searching in a rectangle around the robot given by $\pm 8\,\text{m}$ in x, $\pm 8\,\text{m}$ in y, and $\pm 1\,\text{rad}$ in orientation. We only incorporated links between poses with an information gain above $\gamma = 9\,\text{nats}$.

Figure 4.8 shows the shortest and minimum-uncertainty paths between the chosen start and goal poses, with the corresponding accumulated costs shown in Fig. 4.9.

With this dataset, full matrix recovery is not feasible with the computing resources used, and column-wise marginal computation is impractically slow. Therefore, marginal covariances are approximated using Markov blankets [33]. In this method the marginal covariance for a given pose is approximated considering only the subgraph

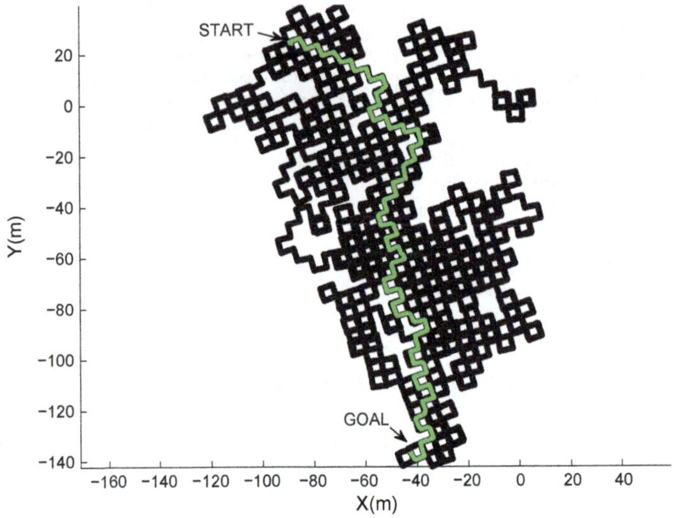

(a) Plan in configuration space

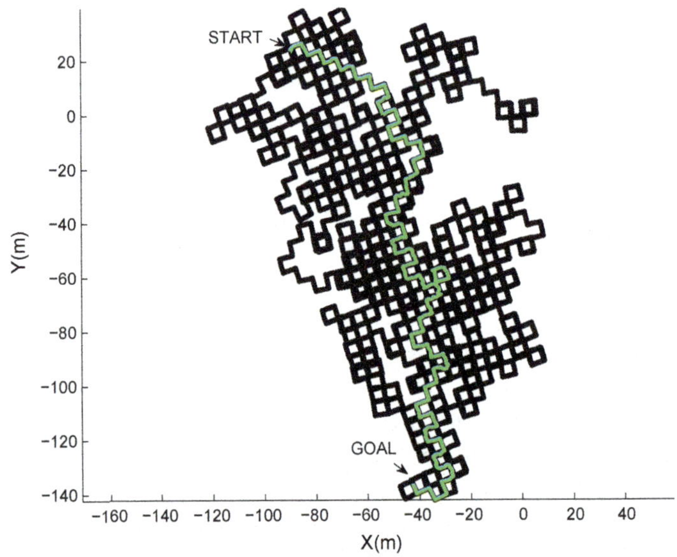

(b) Plan in belief space.

Fig. 4.8 Path planning over the Manhattan dataset. **a** Planning in configuration space we obtain the shortest path to the goal on the underlying Pose SLAM graph. **b** Planning in belief space we obtain the minimum uncertainty path to the goal

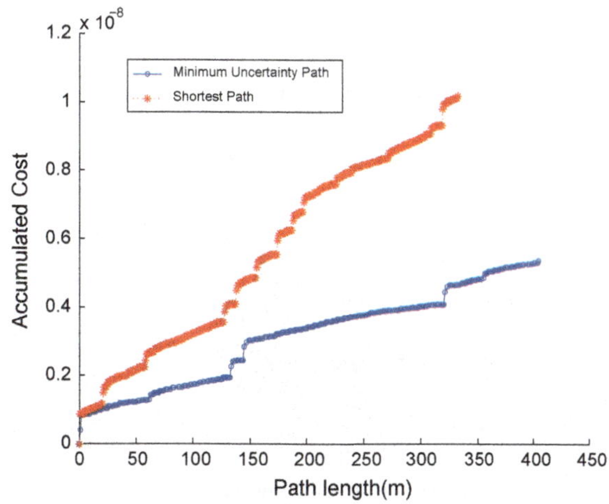

Fig. 4.9 Accumulated cost along the shortest (*red*) and minimum uncertainty (*blue*) path in the Manhattan experiment

of poses directly connected to it, which is typically small. As expected, the cost of the minimum uncertainty path obtained using Markov blankets is significantly better than that of shortest path, with only a marginal increase in path length. The planning time in this case is 122 s, which is reasonable considering that the planner was implemented in Matlab and that the problem includes more than 10000 poses. Thus, even when computing resources are a constraint, the presented method can still be used to plan a route for the robot to the best sensor registration regions at the expense of a possibly degradation in the quality of the final path.

To analyze the effect of using approximated marginal covariances, the experiment with the Manhattan dataset was repeated, but this time using only a subset with the first 2700 poses, only to be able to compare the Markov blanket approximation with the computation of exact covariances. Figure 4.10 shows path planning results over this section of the Manhattan dataset. The use of the Markov blankets reduces the planning time by 50% but it hardly changes the obtained path, validating the approximation. Figure 4.11 shows the accumulated cost in this experiment. As expected, the path length and cost when using the Markov blanket approximation are a compromise between the ones obtained with exact covariances and those of the shortest path.

4.3.4 Dense 3D Mapping Dataset

So far, we have showed results over two dimensional datasets. In this experiment we show the application of our approach to plan a minimum uncertainty escape route on a dense 3D map using range data, as shown in Fig. 4.12.

The experimental data was acquired at the interior plaza of the FME building at UPC, which encompasses a 100 × 40 sqm. rectangular area with various terrain types (gravel, earth, grass) and ramps. The robot used is Teo, a Segway RMP 400

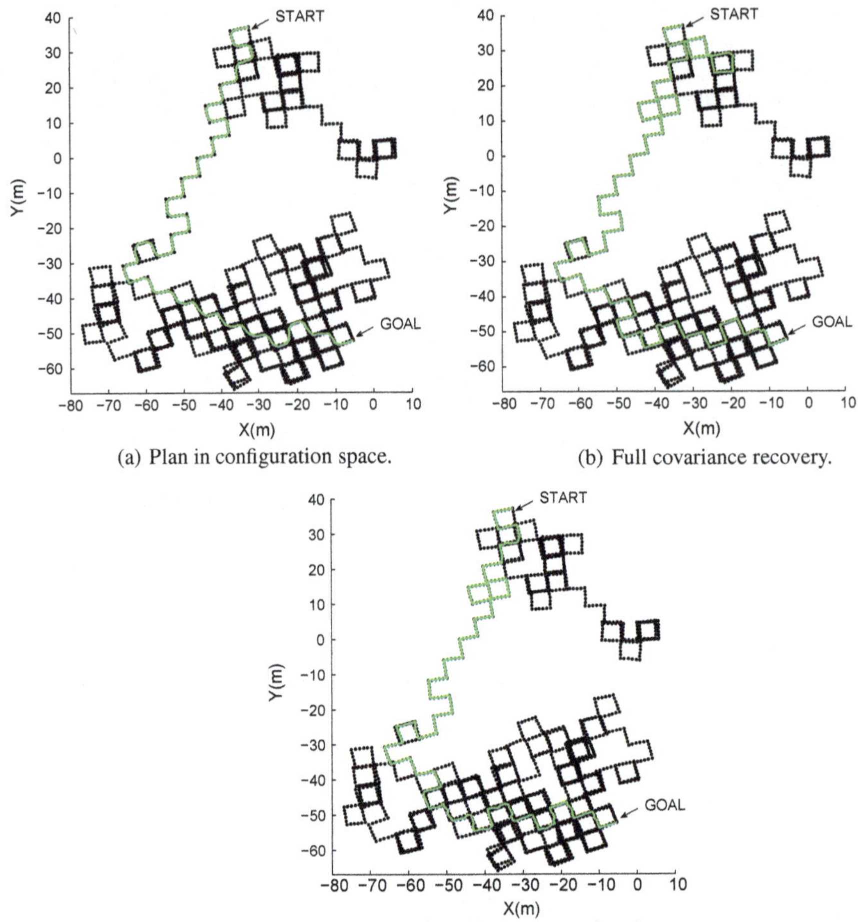

(a) Plan in configuration space.

(b) Full covariance recovery.

(c) Markov blanket approximation.

Fig. 4.10 Path planning over a section of the Manhattan dataset. **a** Planning in configuration space we obtain the shortest path to the goal on the underlying Pose SLAM graph. **b** Planning in belief space we obtain the minimum uncertainty path to the goal. **c** A minimum uncertainty path to the goal computed when the marginal covariances are recovered with Markov blankets

platform equipped with a custom built 3D scanner with a Hokuyo UTM-30LX sensor mounted on a slip ring (see Fig. 4.13). Each aggregated laser scan has 194, 500 points with resolutions of 0.5 deg azimuth and 0.25 deg elevation and range of 30 m, with a noise level of 5 cm in depth. The Pose SLAM map built contains 30 dense point clouds with a maximum separation between consecutive poses of 18 m.

Sensor registration is computed by aligning the range scans with hierarchical ICP [37]. The point clouds were subsampled uniformly using a voxel size of 35 cm and noise was removed using a density policy.

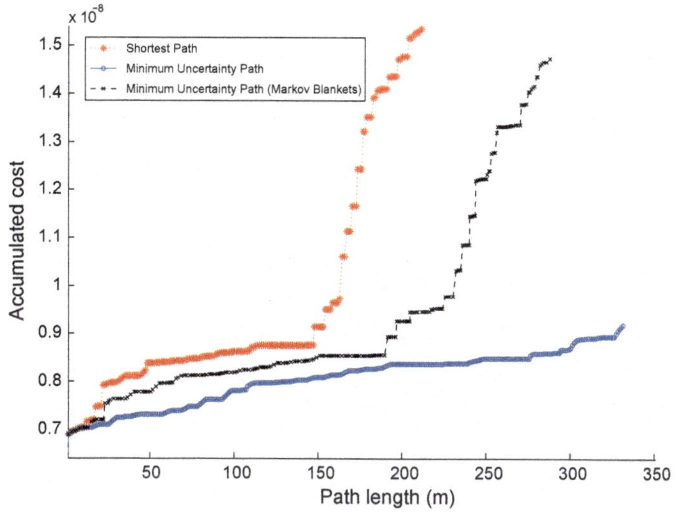

Fig. 4.11 Accumulated cost along the shortest (*red*) and along the minimum uncertainty path computed with exact marginal covariances (*blue*) and with Markov blankets (*black*)

Fig. 4.12 A close in on the computed 3D range map and the robot trajectory

Sensor covariance is approximated with first order error propagation by computing the implicit function Jacobian for ICP's point-to-point unconstrained minimization as shown in [38]. Two factors make this computation suboptimal. On the one hand, it is only a first order approximation, thus conservative. On the second hand it is formulated only for the point-to-point error metric, whilst the ICP implementation is optimized for performance with a hierarchical structure that uses a point-to-plane

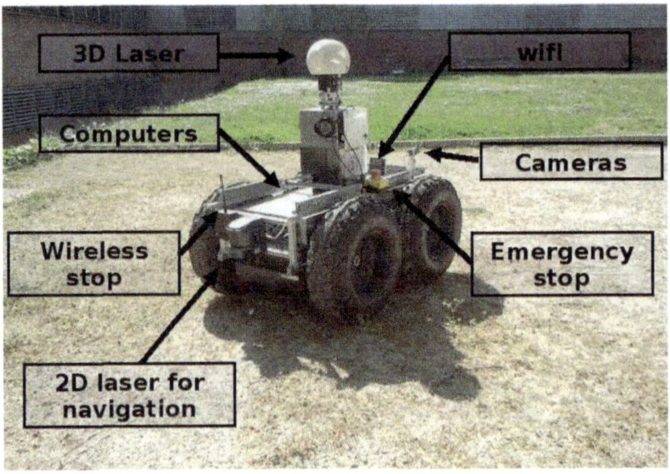

Fig. 4.13 Segway RMP 400 robotic platform at the FME plaza

error metric at the coarsest level and a point-to-point metric at finer levels, and that weights differently rotations and translations [37]. Our experiments have shown empirically that the computation of $\boldsymbol{\Sigma}_y$ is accurate enough and does not jeopardize the rest of the method.

For this experiment we employ our 6 DOF Pose SLAM implementation with Euler angles, described in Chap. 3. The resulting Pose SLAM is shown in Fig. 4.14. A 2D projection of the 3D pose graph is shown in Fig. 4.15a-b. The map contains one situation for which the displacement is so large it precludes sensor registration. For that case, the link in the graph was updated purely with platform odometry data and constant noise covariance $\boldsymbol{\Sigma}_u = \text{diag}(0.0158\,\text{m}, 0.0158\,\text{m}, 0.0791\,\text{m}, 0.0028\,\text{rad}, 0.0028\,\text{rad}, 0.0001\,\text{rad})^2$. The covariance of the initial pose was set to $\boldsymbol{\Sigma}_0 = \text{diag}(0.01\,\text{m}, 0.01\,\text{m}, 0.01\,\text{m}, 0.0087\,\text{rad}, 0.0087\,\text{rad}, 0.0087\,\text{rad})^2$.

During path planning, neighboring poses are linked for a threshold of $\pm 5\,\text{m}$ in x and y and no orientation restriction, thanks to the omnidirectional characteristic of our range sensor. Path search is performed over a 2D projection of the 3D pose graph, marginalizing the x, y and θ variables from the full state vector and state covariance for the computation of the cost function and other path-planning related routines.

The task at hand is to plan a minimum uncertainty escape route from the center of the plaza to the exit of the building. A plan in configuration space finds the shortest path to the goal (see Fig. 4.15a). This route is about 130 m long, but had the drawback of having highly correlated localization uncertainty along the x and y directions from the very beginning, as shown by the projected hyperellipsoids of equiuncertainty. Taking this route to escape could mean the robot getting lost.

A safer route is a path searched in belief space. The plan is a little longer, about 160 m, but with higher guarantee of good sensor registration during path execution, and hence good localization estimates throughout the trajectory (see Fig. 4.15b).

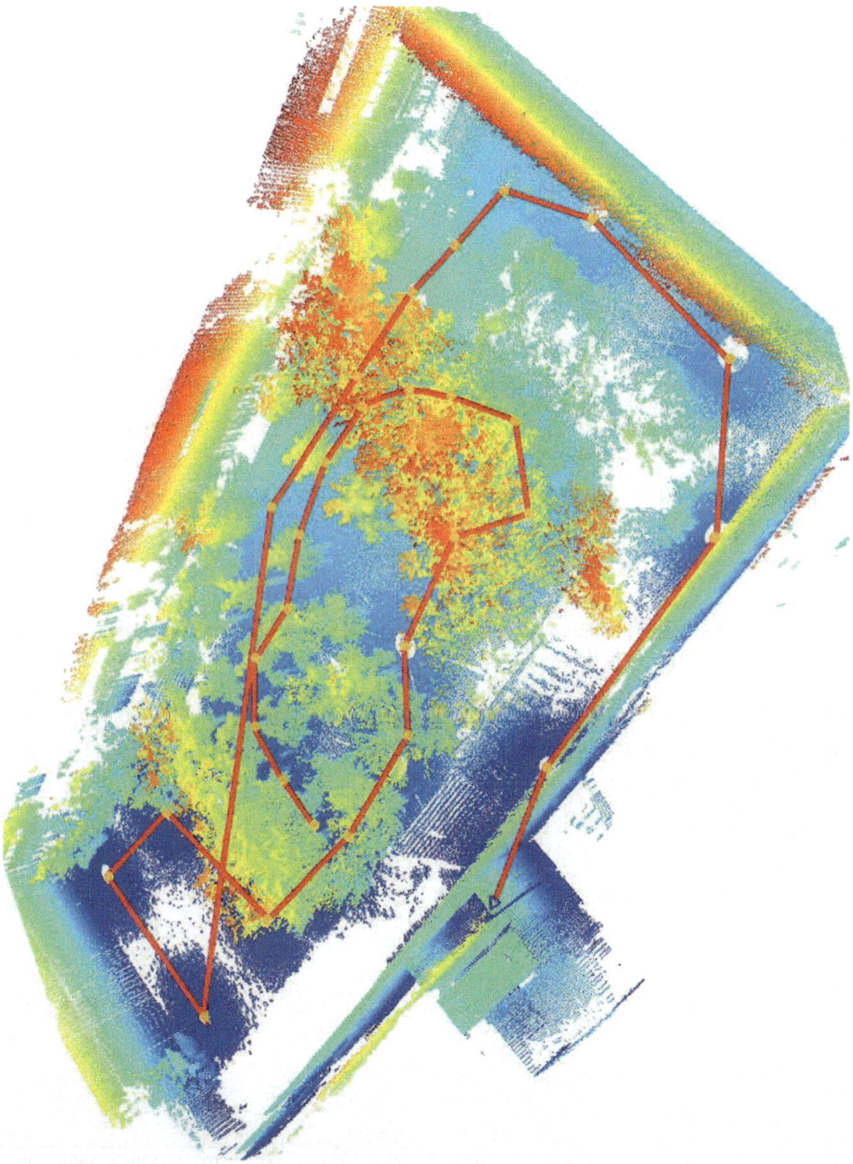

Fig. 4.14 3D Pose SLAM map of the FME plaza, with the robot trajectory shown in *red*

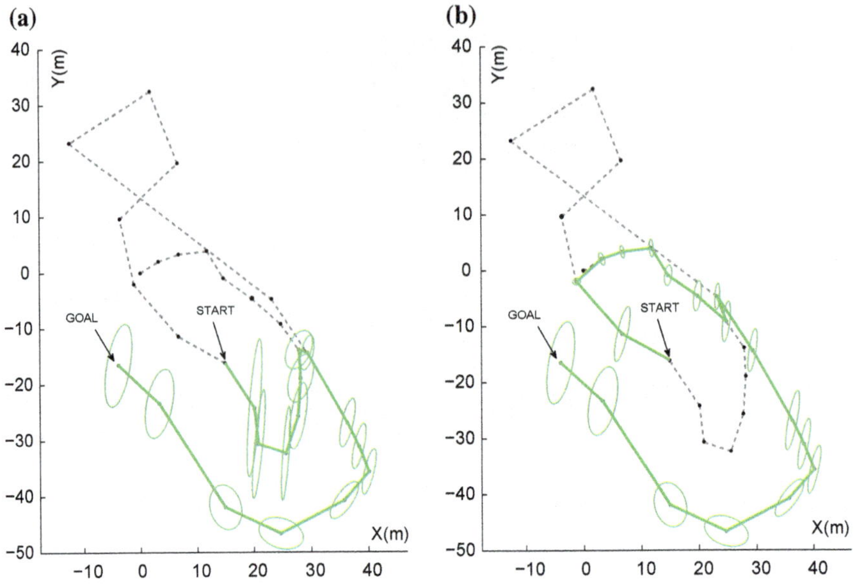

Fig. 4.15 **a** Planning in configuration space we obtain the shortest path to the goal and related covariances. **b** Planning in belief space we obtain the minimum uncertainty path to the goal

Fig. 4.16 Accumulated cost along the shortest path (*red*) and the minimum uncertainty path (*blue*)

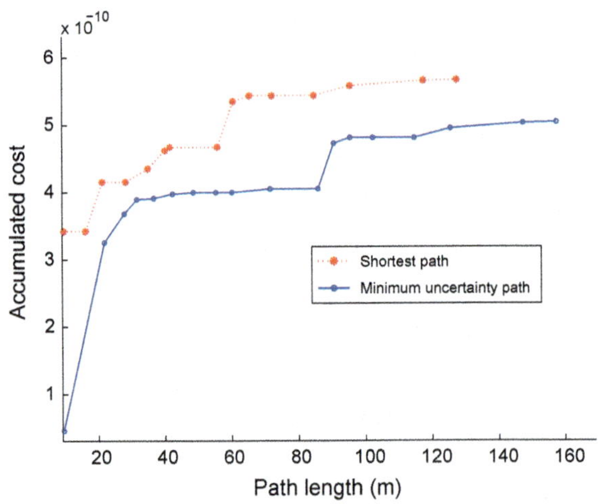

The plot in Fig. 4.16 compares the cost of executing both the shortest path and the minimum uncertainty path as well as the corresponding path lengths.

4.3.5 Real Robot Navigation

To validate the planner in realistic conditions, in this experiment we performed autonomous navigation with our Segway RMP 400 robotic platform (see Fig. 4.13) in an outdoor scenario with uneven and sandy terrain. This is the very same scenario from the experiments shown in Sect. 4.3.4. However, in this experiment our goal is to actually execute the path that our method computes. Compared to the experiment shown in Sect. 4.3.4, a further complexity is added to this experiment by mapping and navigating using 2D laser scan data, instead of the richer 3D dense scan data.

We first acquired data to build a Pose SLAM map using dead-reckoning readings and laser scans over 350 m. The laser readings were used to assert loop closures by aligning them using an ICP algorithm. The Segway dead reckoning readings and the laser pose constraints were modelled with noise covariances $\boldsymbol{\Sigma}_u = \mathrm{diag}(0.0316\,\mathrm{m}, 0.0158\,\mathrm{m}, 0.1104\,\mathrm{rad})^2$, and $\boldsymbol{\Sigma}_z = \mathrm{diag}(0.2\,\mathrm{m}, 0.2\,\mathrm{m}, 0.03\,\mathrm{rad})^2$, respectively, and the uncertainty of the initial pose was set to $\boldsymbol{\Sigma}_0 = \mathrm{diag}(0.1\,\mathrm{m}, 0.1\,\mathrm{m}, 0.09\,\mathrm{rad})^2$. The local planner used was based on the dynamic window approach [39] available in ROS. Figure 4.17 shows the path estimated by Pose SLAM. The red

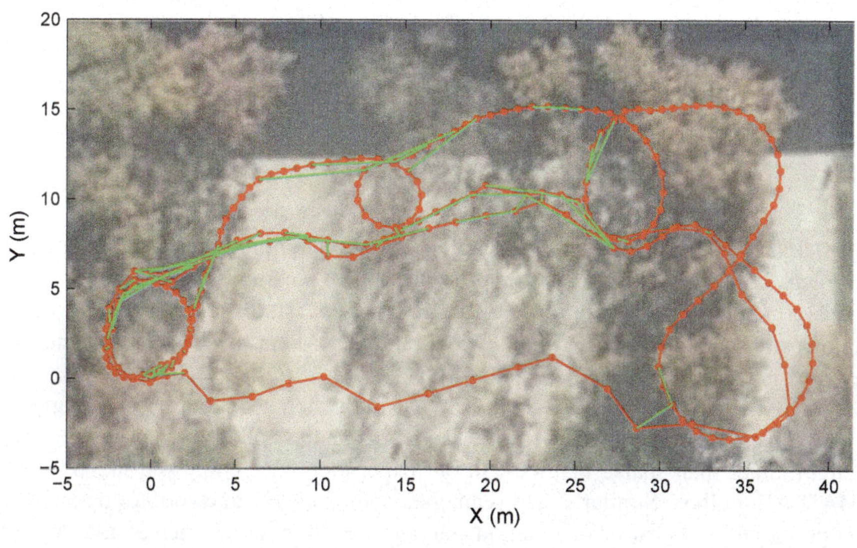

Fig. 4.17 Pose SLAM map built with encoder odometry and laser data in an outdoor scenario with a Segway RMP 400 robotic platform

dots and lines represent the estimated path and the green lines indicate loop closure constraints established by registering scans at non-consecutive poses.

Using this map we computed the shortest and the minimum uncertainty paths that connect two robot configurations on opposite sides of the map, as shown in Fig. 4.18. For the planning, the thresholds to detect nearby poses are set to ± 4.5 m in x, ± 4.5 m in y or ± 1.04 rad in orientation with $s = 0.1$, and the minimum information gain was set to $\gamma = 1.5$ nats.

The shortest path shown in Fig. 4.18a, enters into an uneven and sandy region. The rugged terrain caused the laser to occasionally point to the soil which complicated the registration of the sensor readings. The sand caused some slip that affected wheel odometry. Both effects contributed to produce a patch of the Pose SLAM map with higher uncertainty.

In contrast, the path computed with our approach, shown in Fig. 4.18b, avoids this region. This path traverses an even region of the environment in which the map has lower uncertainty, thanks to the better sensor registration and the more reliable odometry.

Figure 4.19 shows the accumulated cost along the two paths. We can note that the safest path is only 9 m longer than the shortest path. In this example, the total time to compute the plan was 6.5 s, which is significantly smaller than the 12 min required to execute it.

To verify that the assumptions taken in the planning hold in real conditions, we executed both paths with the robot for five times and the obtained trajectories are shown in Fig. 4.20. The result of executing the shortest path are shown in Fig. 4.20a. In this case, the robot was not able to reach the goal for any of the trials. On the contrary, the execution of the safest path, shown in Fig. 4.20b, resulted in the robot safely arriving to the goal in all trials, with an error in the interval of 0.5 m to 1.7 m.

4.3.6 Planning with Graph SLAM Approaches

Our algorithm can also be directly applied to compute minimum uncertainty paths in maps built with other SLAM approaches. Next, we show a comparison of the paths computed with our approach to go from the same starting and ending positions in maps built with Pose SLAM and two Graph SLAM approaches. To this end, we employ g2o[1] [40] and GTSAM[2] [41–43].

GTSAM is a library of C++ classes that implements smoothing and mapping (SAM) using factor graphs and Bayes networks as the underlying computing paradigm rather than sparse matrices. In our experiments we employed the GTSAM MATLAB toolbox. Similarly, g2o is an open-source C++ framework for optimizing graph-based nonlinear error functions, applicable to problems such as SLAM and Bundle Adjustment, which exploits the sparse connectivity of the graph and the spe-

[1] Code accessed from: https://github.com/RainerKuemmerle/g2o.

[2] GTSAM Version 3.2.1 accessed from: https://collab.cc.gatech.edu/borg/gtsam.

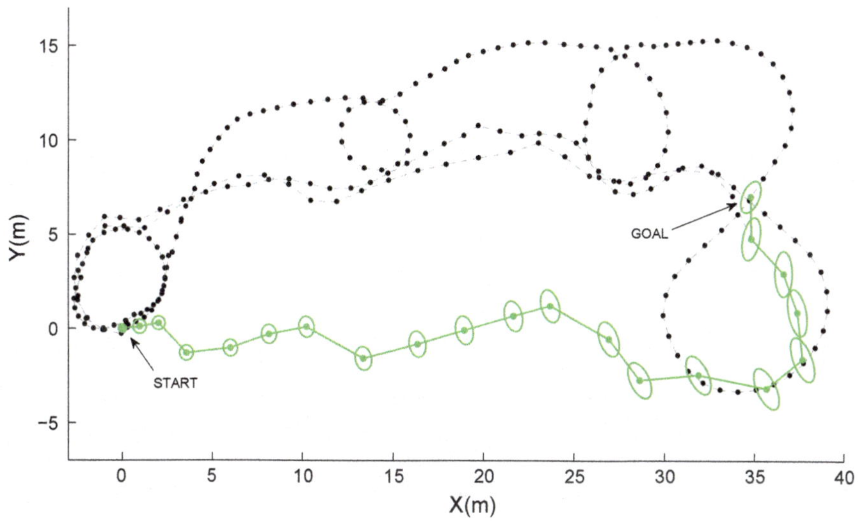

(a) Shortest path to the goal.

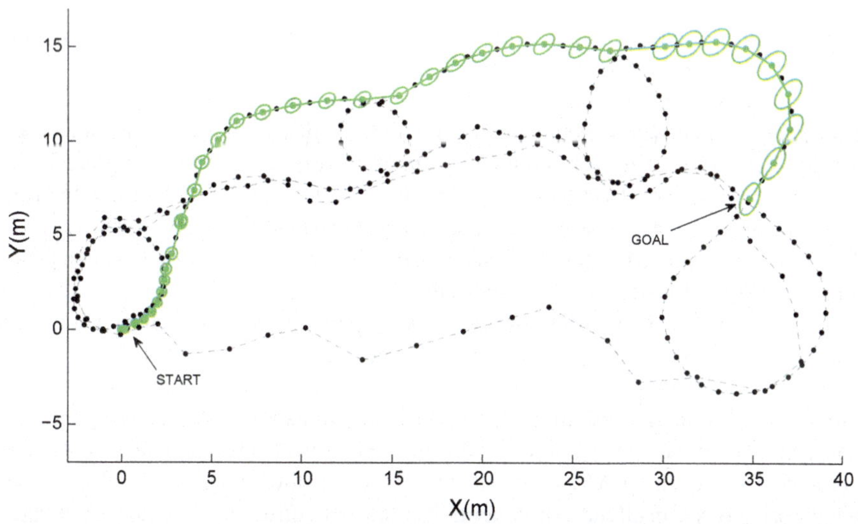

(b) Minimum uncertainty path to the goal.

Fig. 4.18 Path planning over the map built with our mobile robot using encoder odometry and laser data

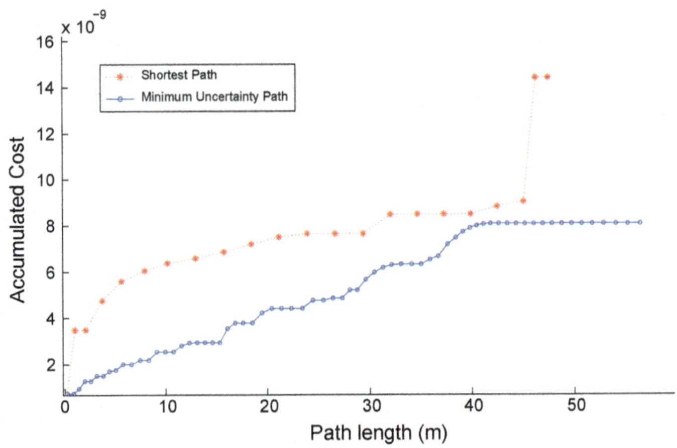

Fig. 4.19 Accumulated cost along the shortest (*red*) and minimum uncertainty (*blue*) path in the real robot experiment

cial structures of the graph that often occur in both applications. In our experiments we employed the default settings in both distributions.

We planned paths in the maps built from the simulated Manhattan data set [36], containing 3500 poses, which is included in the g2o distribution.[3] In such a dataset, the noise covariances for robot odometry and relative-pose measurements are $\boldsymbol{\Sigma}_u = \boldsymbol{\Sigma}_z = \text{diag}(0.1495\,\text{m}, 0.1495\,\text{m}, 0.1495\,\text{rad})^2$. The threshold to detect neighboring poses was set to $s = 0.1$ searching in a rectangle around the robot given by $\pm 8\,\text{m}$ in x, $\pm 8\,\text{m}$ in y, and ± 1 rad in orientation. We only incorporated links between poses with an information gain above $\gamma = 9$ nats.

Figure 4.21 shows the minimum-uncertainty paths between the chosen start and goal poses, with the corresponding accumulated costs shown in Fig. 4.22.

From Fig. 4.21 we can notice that the final paths are similar, they all drive the robot throughout practically the same regions. The distinctions between the three minimum uncertainty paths can be explained by the differences in the estimated pose marginals obtained from each SLAM approach and by using the same threshold to detect neighboring poses in all the experiments. That is, with different covariance estimates distinct nodes were connected in the path search. The candidate neighboring nodes dependency on the covariance estimates is described in Sect. 4.1.1. The difference in the covariance size is also evident in the final accumulated costs shown in Fig. 4.22 as we compute the step uncertainty using the marginal covariance for the target pose as described in Sect. 4.1.2.

[3] Accessed from: https://svn.openslam.org/data/svn/g2o/trunk/data/2d/manhattan3500/.

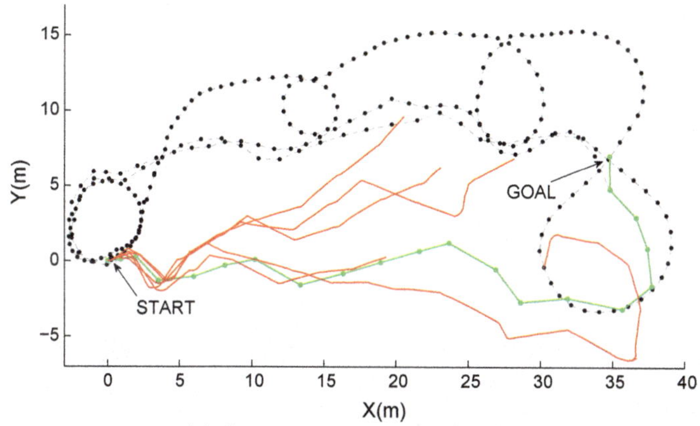

(a) 5 attempts to execute the shortest path.

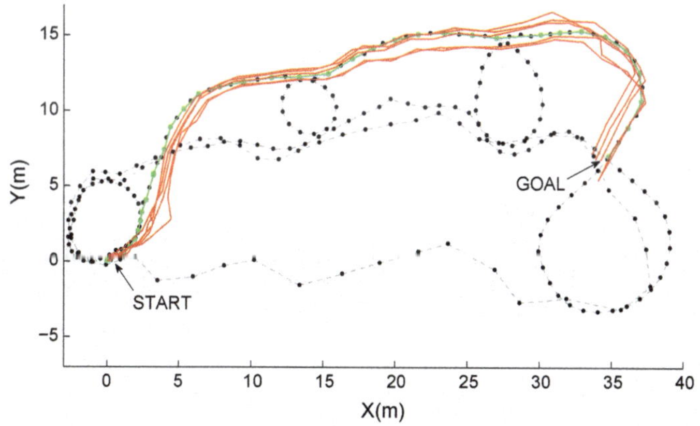

(b) 5 executions of the safest path.

Fig. 4.20 Real path execution of the shortest and safest paths to the goal with our mobile robot. The *green line* shows the planned paths computed with our method. The *red lines* represent the obtained trajectories when executing each path five times. The execution is interrupted when the deviation with respect to the intended plan is above a safety threshold

The main reason of obtaining such differences in marginal covariances are due to the way the graph is optimized in each SLAM method and the way that each marginal is recovered. Generally speaking, in Graph SLAM approaches (i.e. g2o and GTSAM in our experiment), at every iteration, the previous solution is used as the linearization point for the Jacobians, while in Pose SLAM a single point is used, thus obtaining less refined covariance estimates. In any case, although we got different uncertainty values in the three cases, such a difference is roughly uniform in all the pose estimates, i.e. the regions with high and low uncertainty remain approximately the same in the three maps, thus obtaining similar low uncertainty paths.

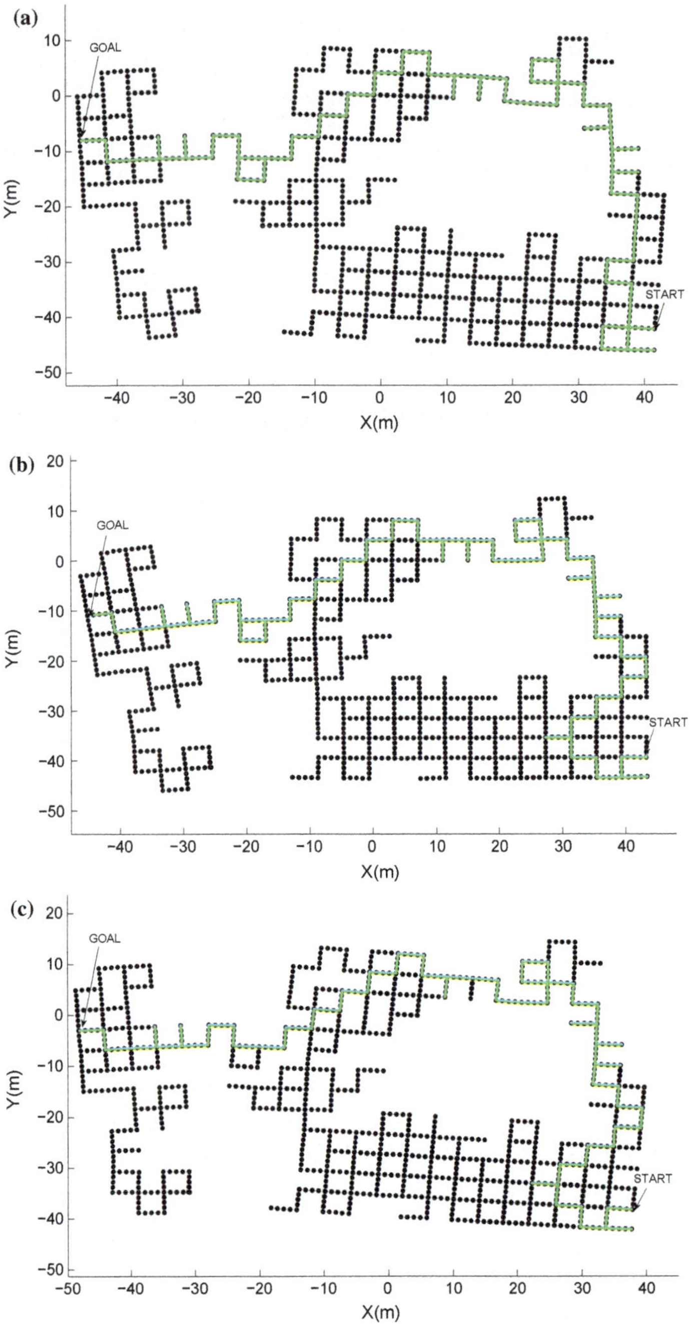

Fig. 4.21 Planning in belief space with **a** Pose SLAM, **b** GTSAM and **c** g2o

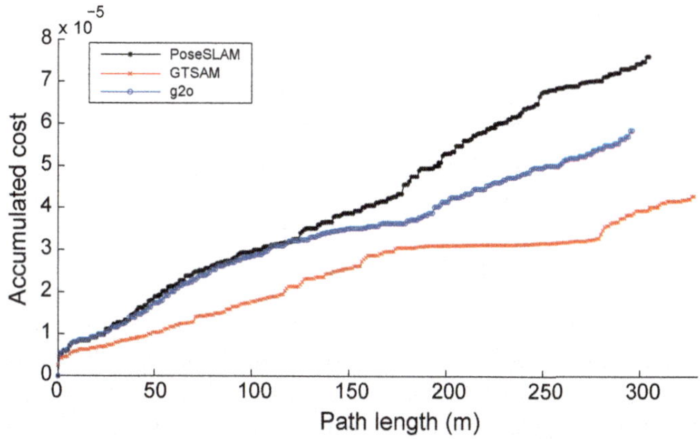

Fig. 4.22 Accumulated cost along the minimum uncertainty path using Pose SLAM (*black*), GTSAM (*red*) and g2o (*blue*)

In conclusion, we can note that our approach can be directly applied to compute minimum uncertainty paths in maps built with other SLAM approaches. Although maps of the same environment built with different SLAM methods (even using the same measurements) may differ, our path planning technique can be used to drive the robot to the lowest uncertainty regions in such maps, where we can expect it to be better localized during path execution.

4.4 Bibliographical Notes

The work presented in this chapter is in the intersection of three disciplines: planning under uncertainty, motion planning, and SLAM.

Partial Observable Markov Decision Processes (POMDP) provide the most general framework for planning under uncertainty. In a POMDP, the knowledge about the agent's state is encoded as a *belief* (a probability distribution over all possible states) and the objective is to determine a *policy* giving the best action for each belief. Robot navigation is naturally modeled as a continuous problem but, unfortunately there are very few approaches able to deal with POMDPs in continuous spaces [44, 45]. Thus, the usual approach discretizes the problem and applies standard value iteration algorithms [46, 47]. Unfortunately, those algorithms can only deal with problems of low dimensionality. Point-value iteration algorithms [48, 49] somehow alleviate these problems focusing the planning to the reachable belief space. However, they are not efficient enough to be applied to large-scale navigation problems. Another inconvenient of standard POMDP algorithms is that they assume a known model of the environment.

Motion planning [50] deals with the problem of finding adequate trajectories to reach distant locations. For low dimensional spaces approximate cell decomposition provides solutions to the motion planning process by discretizing the environment in cells, selecting obstacle-free ones and finding a shortest path to the goal using a standard shortest path algorithm. However, for higher dimensional spaces the approach becomes unfeasible due to the curse of dimensionality. The solution is to resort to a roadmap-based method. A roadmap is a collection of one-dimensional curves that capture the topology of the configuration space. Paths between two given configurations are obtained traveling over the roadmap. The silhouette method [51] defines roadmaps with a guarantee of completeness, but can only be applied to small problems. Stochastic variants such as Probabilistic Roadmap (PRMs) or Rapidly Expanding RandomTrees (RRTs) only offer probabilistic completeness, but can be successfully applied to problems with high dimensionality [13, 14]. The main issues with classical motion planning algorithms when applied to navigation is that they assume a known model of the environment and that they do not take into account the inherent noise in robot motion.

Several combinations of planning under uncertainty, SLAM, and motion planning exist. Planning under uncertainty and mapping are combined in approaches that attempt to simultaneously capture the environment model and optimize the policy [52, 53]. Up to now, those approaches are only valid for relatively small problems. Motion planning taking into account uncertainty is also present in contributions that consider the noise in the environment model [17] or in the robot pose due to the uncertain effects of actions [1, 18, 19, 54, 55]. Motion planning algorithms incorporating cost functions [56, 57] can also accommodate uncertainty.

Some approaches have attempted to perform motion planning together with SLAM. However, the classical representation of the environment in SLAM hampers their performance. Initial work in SLAM represented the environment using a sparse set of features, this type of representation needs to be enriched with obstacles or traversability related information before it can be used for collision-free path planning. For instance, in [10], the map of the environment is enriched with virtual free-space markers connected defining a graph of traversable regions. A robot would navigate first to the nearest free-space marker and then follow the free-space graph to the goal. Alternatively, the hybrid metric maps (HYMMs) in [11] split the environment in local triangular regions (LTR) whose corners are features in the map. Each LTR has local paths that traverse it and an associated cost for each of these local paths so that a planner can find the lowest cost path from any point to the goal through a sequence of LTRs. HYMMs were improved to include a scalar evaluation of the potential information that the robot can obtain from the environment at each LTR [12]. This is relevant information for path planning, but comes at the expense of significant increase in complexity and in memory use.

Instead of enriching feature-based maps, other approaches build grid maps out of volumetric representations of 3D point clouds [58, 59]. These techniques, typically use the 3D map to extract a 2D map of traversable regions from which a graph-like roadmap is derived. Such graph is then used for path planning relying on standard graph search algorithms. Those approaches, however, also come at the expense of increased complexity and higher memory cost. Moreover, traversability is typically computed over the mean estimate of the map, disregarding map uncertainty.

An alternative is to compute a lighter representation of the environment such as a topological map [60–62] and to use it for path planning [63]. However, since topological maps are not accurate enough to localize the robot in all cases, they are sometimes combined with local grid maps [64–66]. In these approaches, the topological map is used as a roadmap to devise a path to the goal using graph search techniques, and the local grid map associated with each node in the topological map is used to compute a path from one node to the next, considering local obstacle avoidance and path smoothing. Although the use of hybrid metric-topological maps improves the efficiency of path planning when compared to the use of a global grid map, these approaches still require considerable effort to maintain the coherence between both representations, especially at loop closure. As with the traversability maps, the computed routes on topological maps also ignore the sources of uncertainty included in the map.

In this Chapter we observe that the maps computed with Pose SLAM can be directly used as belief roadmaps and, thus, used for planning low uncertainty paths without further processing the map nor enriching it with additional information. Moreover, since we rely on Pose SLAM which marginalizes out the sensor readings, the approach can be used with any type of sensors. In [9] it is suggested to use the graph of poses built with bundle adjustment for path planning, but the uncertainty information in the map is not exploited in the computation of the optimal path. In contrast, we use the maps computed with Pose SLAM to plan in the belief space obtaining paths to remote locations that take into account the uncertainty along them.

The approach is devised to autonomously guide the robot in scenarios where the robot had already built a map. This mapping session need not be exhaustive as long as it traverses all areas the robot is intended to visit in normal operations. Note however that the technique could also be used to plan routes in a partially built scenario during autonomous exploration. In [67] for instance, map optimization and pose optimization are jointly taken into account during exploration by defining a set of visibility nodes using a skeleton of the current map free zones, and planning an optimal path through these nodes. The same problem can also be addressed using frontiers instead as the driving nodes for exploration [68].

References

1. S. Prentice, N. Roy, The Belief Roadmap: efficient planning in belief space by factoring the covariance. Int. J. Robot. Res. **29**(11–12), 1448–1465 (2009)
2. R. He, S. Prentice, N. Roy, Planning in information space for a quadrotor helicopter in a GPS-denied environment, in *Proceedings of IEEE International Conference Robotics Automation* (Pasadena, May 2008), pp. 1814–1820
3. R.M. Eustice, H. Singh, J.J. Leonard, M.R. Walter, Visually mapping the RMS Titanic: conservative covariance estimates for SLAM information filters. Int. J. Robot. Res. **25**(12), 1223–1242 (2006)
4. S. Thrun, S. Thayer, W. Whittaker, C. Baker, W. Burgard, D. Ferguson, D. Hahnel, M. Montemerlo, A. Morris, Z. Omohundro, C. Reverte, Autonomous exploration and mapping of abandoned mines. Robot. Automat. Mag. **11**(4), 79–91 (2004)
5. R. Smith, M. Self, P. Cheeseman, A stochastic map for uncertain spatial relationships, in *Proceedings of 4th International Symposium Robotics Research* (Santa Clara, 1988), pp. 467–474
6. K. Konolige, Large-scale map-making, in *Proceedings of 19th AAAI Conference on Artificial Intelligence* (San Jose, California, Jul 2004), pp. 457–463
7. S. Thrun, M. Montemerlo, The graph SLAM algorithm with applications to large-scale mapping of urban structures. Int. J. Robot. Res. **25**(5–6), 403–429 (2006)
8. U. Frese, L. Schröder, Closing a million-landmarks loop, in *Proceedings of IEEE/RSJ International Conference on Intelligent Robots System* (Beijing, Oct 2006), pp. 5032–5039
9. G. Sibley, C. Mei, I. Reid, P. Newman, Vast-scale outdoor navigation using adaptive relative bundle adjustment. Int. J. Robot. Res. **29**(8), 958–980 (2010)
10. P.M. Newman, J. Leonard, J. Neira, J.D. Tardós, Explore and return: experimental validation of real time concurrent mapping and localization, in *Proceedings of IEEE International Conference on Robotics Automation* (Washington, May 2002), pp. 1802–1809
11. J. Guivant, E. Nebot, J. Nieto, F. Masson, Navigation and mapping in large unstructured environments. Int. J. Robot. Res. **23**(4–5), 449–472 (2004)
12. S. Rezaei, J. Guivant, J. Nieto, E. Nebot. Simultaneous information and global motion analysis ("SIGMA") for car-like robots, in *Proceedings of IEEE International Conference Robotics Automation* (New Orleans, Apr 2004), pp. 1939–1944
13. L. Kavraki, P. Svestkaand, J.C. Latombe, M. Overmars, Probabilistic roadmaps for path planning in high dimensional configuration spaces. IEEE Trans. Robot. **12**(4), 566–580 (1996)
14. S. LaValle, J. Kuffner, Randomized kinodynamic planning. Int. J. Robot. Res. **20**(5), 378–400 (2001)
15. N. Ratliff, M. Zucker, J.A. Bagnell, S. Srinivasa, CHOMP: Gradient optimization techniques for efficient motion planning, in *Proceedings of IEEE International Conference on Robotics Automation* (Kobe, May 2009), pp. 489–494
16. S. Karaman, E. Frazzoli, Sampling-based algorithms for optimal motion planning. Int. J. Robot. Res. **30**(7), 846–894 (2011)
17. P. Missiuro, N. Roy, Adapting probabilistic roadmaps to handle uncertain maps, in *Proceedings of IEEE International Conference on Robotics Automation* (Orlando, May 2006), pp. 1261–1267
18. R. Pepy, M. Kieffer, E. Walter, Reliable robust path planner, in *Proceedings of IEEE/RSJ International Conference on Intelligent Robots and Systems* (Nice, Sep 2008), pp. 1655–1660
19. R. Alterovitz, T. Simeon, K. Goldberg, The Stochastic motion roadmap: a sampling framework for planning with Markov motion uncertainty, *Robotics Science and Systems*, vol. 3 (Atlanta, 2007)
20. A. Bry, N. Roy, Rapidly-exploring Random Belief Trees for motion planning under uncertainty, in *Proceedings of IEEE International Conference on Robotics Automation* (Shanghai, May 2011), pp. 723–730
21. V. Ila, J.M. Porta, J. Andrade-Cetto, Information-based compact Pose SLAM. IEEE Trans. Robot. **26**(1), 78–93 (2010)

22. F. Lu, E. Milios, Globally consistent range scan alignment for environment mapping. Auton. Robot. **4**(4), 333–349 (1997)
23. R.M. Eustice, H. Singh, J.J. Leonard, Exactly sparse delayed-state filters for view-based SLAM. IEEE Trans. Robot. **22**(6), 1100–1114 (2006)
24. K. Konolige, M. Agrawal, FrameSLAM: from bundle adjustment to realtime visual mapping. IEEE Trans. Robot. **24**(5), 1066–1077 (2008)
25. N. Michael, E. Stump, K. Mohta. Persistent surveillance with a team of MAVs, in *Proceedings of IEEE/RSJ International Conference on Intelligent Robots and Systems* (San Francisco, Sep 2011), pp. 2708–2714
26. A. Zini, Robots expand delivery options. *Health Management Technology* (March 2011), pp. 10–12
27. R. Platt, R. Tedrake, L. Kaelbling, T. Lozano-Perez, Belief space planning assuming maximum likelihood observations, *Robotics Science and Systems*, vol. 6 (Zaragoza, Spain, 2010)
28. R. Valencia, J. Andrade-Cetto, J.M. Porta, Path planning in belief space with Pose SLAM, in *Proceedings of IEEE International Conference on Robotics Automation* (Shanghai, May 2011), pp. 78–83
29. L. Jaillet, J. Cortés, T. Siméon, Sampling-based path planning on configuration-space costmaps. IEEE Trans. Robot. **26**(4), 635–645 (2010)
30. R. Sim, N. Roy, Global A-optimal robot exploration in SLAM, in *Proceedings of IEEE International Conference on Robotics Automation* (Barcelona, Apr 2005), pp. 661–666
31. V. Ila, J.M. Porta, J. Andrade-Cetto, Amortized constant time state estimation in Pose SLAM and hierarchical SLAM using a mixed Kalman-information filter. Robot. Auton. Syst. **59**(5), 310–318 (2011)
32. Y. Chen, T.A. Davis, W.W. Hager, S. Rajamanickam, Algorithm 887: CHOLMOD, supernodal sparse Cholesky factorization and update/downdate. ACM T. Math. Softw. **35**(3), 22:1–22:14 (2008)
33. S. Thrun, Y. Liu, D. Koller, A.Y. Ng, Z. Ghahramani, H. Durrant-Whyte, Simultaneous localization and mapping with sparse extended information filters. Int. J. Robot. Res. **23**(7–8), 693–716 (2004)
34. M. Quigley, B. Gerkey, K. Conley, T. Foote J. Faust, J. Leibs, E. Berger, R. Wheeler, A.Y. Ng, ROS: an open-source robot operating system, in *Proceedings of IEEE ICRA Workshop Open Source Software Robotics* (Kobe, 2009)
35. A. Howard, N. Roy, The robotics data set repository (Radish) (2003), http://radish.sourceforge.net
36. E. Olson, J. Leonard, S. Teller, Fast iterative alignment of pose graphs with poor initial estimates, in *Proceedings of IEEE International Conference on Robotics Automation* (Orlando, May 2006), pp. 2262–2269
37. E.H. Teniente, J. Andrade-Cetto, FaMSA: fast multi-scan alignment with partially known correspondences, in *Proceedings of European Conference on Mobile Robotics* (Orebro, Sep 2011), pp. 139–144
38. S. Foix, G. Alenyà, J. Andrade-Cetto, C. Torras, Object modeling using a ToF camera under an uncertainty reduction approach, in *Proceedings of IEEE International Conference on Robotics Automation* (Anchorage, May 2010), pp. 1306–1312
39. D. Fox, W. Burgard, S. Thrun, The dynamic window approach to collision avoidance. Robot. Autom. Mag. **4**(1), 23–33 (1997)
40. R. Kummerle, G. Grisetti, H. Strasdat, K. Konolige, W. Burgard, g2o:a general framework for graph optimization, in *Proceedings of IEEE International Conference on Robotics Automation* (Shanghai, May 2011), pp. 3607–3613
41. F. Dellaert, M. Kaess, Square root SAM: simultaneous localization and mapping via square root information smoothing. Int. J. Robot. Res. **25**(12), 1181–1204 (2006)
42. M. Kaess, A. Ranganathan, F. Dellaert, iSAM: Incremental smoothing and mapping. IEEE Trans. Robot. **24**(6), 1365–1378 (2008)
43. M. Kaess, H. Johannsson, R. Roberts, V. Ila, J. Leonard, F. Dellaert, iSAM2: Incremental smoothing and mapping using the bayes tree. Int. J. Robot. Res. **31**(2), 216–235 (2012)

44. S. Thrun, Monte Carlo POMDPs, in *Advances in Neural Information Processing Systems 12* (The MIT Press, Apr 1999), pp. 1064–1070
45. J.M. Porta, N. Vlassis, M.T.J. Spaan, P. Poupart, Point-based value iteration for continuous POMDPs. J. Mach. Learn. Res. **7**, 2329–2367 (2006)
46. E.J. Sondik, The optimal control of partially observable Markov processes over the infinite horizon: discounted cost. Oper. Res. **26**, 282–304 (1978)
47. L.P. Kaelbling, M.L. Littman, A.R. Cassandra, Planning and acting in partially observable stochastic domains. Artif. Intell. **101**, 99–134 (1995)
48. J. Pineau, G. Gordon, S. Thrun, Point-based value iteration: an anytime algorithm for POMDPs, in *Proceedings of International Joint Conference Artificial Intelligent* (Acapulco, 2003)
49. M.T.J. Spaan, N.A. Vlassis, Perseus: randomized point-based value iteration for POMDPs. J. Artif. Intell. Res. **24**, 195–220 (2005)
50. J.-C. Latombe, *Robot Motion Planning* (Kluwer Academic, London, 1991)
51. J.F. Canny, *The Complexity of Robot Motion Planning* (MIT Press, 1988)
52. S. Ross, B. Chaib-Draa, J. Pineau, Bayesian reinforcement learning in continuous POMDPs with application to robot navigation, in *Proceedings of IEEE International Conference on Robotics Automation* (Pasadena, May 2008), pp. 2845–2851
53. P. Dallaire, C. Besse, S. Ross, B. Chaib-draa, Bayessian reinforcement learning in continuous POMDPs with Gaussian processes, in *Proceedings of IEEE/RSJ International Conference on Intelligent Robots and Systems* (Saint Louis, Oct 2009), pp. 2604–2609
54. N. Roy, S. Thrun, Coastal navigation with mobile robots, in *Advances in Neural Information Processing Systems 12* (The MIT Press, Apr 1999), pp. 1043–1049
55. J.P. Gonzalez, A. Stentz, Planning with uncertainty in position using high-resolution maps, in *Proceedings of IEEE International Conference on Robotics Automation* (Rome, Apr 2007), pp. 1015–1022
56. J. Lee, C. Pippin, T. Balch, Cost based planning with RRT in outdoor environments, in *Proceedings of IEEE/RSJ International Conference on Intelligent Robots and Systems* (Nice, Sep 2008), pp. 684–689
57. L. Jaillet, J. Cortés, T. Siméon, Transition-based RRT for path planning in continuous cost spaces, in *Proceedings of IEEE/RSJ International Conference on Intelligent Robots and Systems* (Nice, Sep 2008), pp. 2145–2150
58. R. Kummerle, D. Hahnel, D. Dolgov, S. Thrun, W. Burgard, Autonomous driving in a multi-level parking structure, in *Proceedings of IEEE International Conference on Robotics Automation* (Kobe, May 2009), pp. 3395–3400
59. D. Dolgov, S. Thrun, Autonomous driving in semi-structured environments: mapping and planning, in *Proceedings of IEEE International Conference on Robotics Automation* (Kobe, May 2009), pp. 3407–3414
60. H. Choset, K. Nagatani, Topological simultaneous localization and mapping (SLAM): toward exact localization without explicit localization. IEEE Trans. Robot. Autom. **17**(2), 125–137 (2001)
61. F. Fraundorfer, C. Engels, D. Nister, Topological mapping, localization and navigation using image collections, in *Proceedings of IEEE/RSJ International Conference on Intelligent Robots and Systems* (San Diego, Nov 2007), pp. 3872–3877
62. T. Goedeme, M. Nuttin, T. Tuytelaars, L. Van Gool, Omnidirectional vision based topological navigation. Int. J. Comput. Vision **74**(3), 219–236 (2007)
63. S. Thrun, Learning metric-topological maps for indoor mobile robot navigation. Artif. Intell. **99**(1), 21–71 (1998)
64. M.A. Whitty, J. Guivant, Efficient path planning in deformable maps, in *Proceedings of IEEE/RSJ International Conference on Intelligent Robots and Systems* (Saint Louis, Oct 2009), pp. 5401–5406
65. M.A. Whitty, J. Guivant, Efficient global path planning during dense map deformation, in *Proceedings of IEEE International Conference on Robotics Automation* (Shanghai, May 2011), pp. 4943–4949

66. K. Konolige, E. Marder-Eppstein, B. Marthi, Navigation in hybrid metric-topological maps, in *Proceedings of IEEE International Conference on Robotics Automation* (Shanghai, May 2011), pp. 3041–3047
67. T. Kollar, N. Roy, Efficient optimization of information-theoretic exploration in SLAM, in *Proceedings of 23th AAAI Conference on Artificial Intelligence* (Chicago, Jul 2008), pp. 1369–1375
68. R. Valencia, J. Valls Miró, G. Dissanayake, J. Andrade-Cetto, Active pose SLAM, in *Proceedings of IEEE/RSJ International Conference on Intelligent Robots and Systems* (Algarve, Oct 2012), pp. 1885–1891

Chapter 5
Active Pose SLAM

This Chapter presents an active exploration strategy that complements Pose SLAM and the path planning approach shown in Chap. 4. This strategy evaluates the utility of exploratory and place revisiting sequences and chooses the one that minimizes overall map and path entropies. The technique considers trajectories of similar path length taking marginal pose uncertainties into account. An advantage of the proposed strategy with respect to competing approaches is that to evaluate information gain over the map, only a very coarse prior map estimate needs to be computed. Its coarseness is independent and does not jeopardize the Pose SLAM estimate. Moreover, a replanning scheme is devised to detect significant localization improvement during path execution.

In spite of the advances in the SLAM problem, most SLAM techniques to date are passive in the sense that the robot only estimates the model of the environment, but without taking any decisions on its trajectory. An active technique on the contrary, would also compute the appropriate robot actions to reduce the uncertainty about its own localization and the map, while at the same time optimizing coverage [1, 2]. A straightforward solution is to combine a classical exploration method with a SLAM technique. However, classical exploration methods focus on reducing the amount of unseen area disregarding the cumulative effect of localization drift, leading the robot to accumulate more and more uncertainty. A solution to the problem should revisit known areas from time to time, trading off coverage with accuracy.

Although action selection is the central issue in exploration for SLAM, there are also other issues that need to be considered. We need to choose a SLAM method, an environment representation, a coverage strategy, and an objective function. Each one imposes different challenges. Besides the well-known challenges in the SLAM problem (e.g. scalability, consistency, data association, robot perception), the selection of the adequate objective function is determinant to define the quality of the map as well as the strategy to cover efficiently the environment. Regarding the action selection, a key challenge is to trade off between optimality and efficiency.

R. Valencia and J. Andrade-Cetto, *Mapping, Planning and Exploration with Pose SLAM*, Springer Tracts in Advanced Robotics 119,
DOI 10.1007/978-3-319-60603-3_5

In this Chapter we tackle the exploration problem for the case of Pose SLAM [3], in order to automate the roadmap construction from scratch by selecting the appropriate actions to drive the robot so as to maximize coverage and at the same time minimize localization and map uncertainties.

To guarantee coverage, an occupancy grid of the environment is maintained. A significant advantage of the approach is that this grid is only computed to hypothesize entropy reduction of candidate map posteriors, and that it can be computed at a very coarse resolution since it is not used to maintain neither the robot localization estimate, nor the structure of the environment. In a similar way to [4], the technique evaluates two types of actions: exploratory actions and place revisiting actions. Action decisions are made based on entropy reduction estimates. By maintaining a Pose SLAM estimate at run time, the technique allows to replan trajectories online should significant change in the Pose SLAM estimate be detected, something that would make the computed entropy reduction estimates obsolete.

The rest of the Chapter is structured as follows. The set of actions is described in Sect. 5.1, and the computation of their utility is described in Sect. 5.2. Replanning is covered in Sects. 5.3, and 5.4 describes a set of experiments that validate the strategy. Finally, in Sect. 5.5 we include a review on the problem of exploration for SLAM.

5.1 Action Set

Evaluating the effect of potential actions in the context of SLAM is expensive since the posterior must be evaluated for each candidate. A continuous set of actions is considered in the active learning approach shown in [5]; however, the learning must be performed per-step basis, that is, for each new destination, a new value function must be learned. On the contrary, in [6] the authors presents a similar approach that aims to learn a general purpose policy instead. Nevertheless, it still relies on a discretization of the state and actions spaces. In any case, these approaches still provide a suboptimal solution and increase the complexity of the exploration strategy. Instead, in this approach we choose to favor scalability by selecting only a limited set of actions as it is done in [4], where actions are defined as trajectories of two types: exploratory sequences and place re-visiting sequences. Next, we show how to compute these actions in the context of Pose SLAM.

5.1.1 Exploratory Actions

Exploratory actions are computed by classical frontier-based exploration [7] and are intended to maximize coverage, that is, to drive the robot towards unknown regions. As in our case we do not have access to an explicit metric representation of the obstacles we have to generate it. However, the necessary data is implicitly encoded in the Pose SLAM posterior. Moreover, as this metric representation is not needed for the estimation process (i.e. the SLAM process), as long as we can find frontiers and plan a path to those frontiers we are free to build this map as coarse as possible.

In the implementation reported here, the Pose SLAM algorithm stores the laser scans corresponding to each of the nodes in the pose graph. It is possible to use these scans to render an occupancy grid [8] for goal selection. Our premise of maximum likelihood navigation suggests that the map will not significantly change during path traversal, but only upon completing a path or should replanning be triggered due to large map shift at loop closure. This situation prevents us from online map rendering, since its computation is only needed at those events in time. Figure 5.1a shows a Pose SLAM map. Frame (b) and (c) in the same figure show the rendered occupancy probability $p(\mathbf{O}_m)$ for the occupancy grid map $\mathbf{O}_m$ at the mean prior, with a resolution of 20×20 cm.

Once we have computed the occupancy grid map, we extract frontiers and plan a path from the last robot pose to reach them. Following [7], frontiers are the set of free cells that have at least one neighboring unknown cell. Once we have identified the set of frontier cells we apply connected component labeling to detect connected regions of frontier cells. Then, for each frontier region of size larger than a threshold, we obtain its center of mass and compute a path from the last robot pose to that point. Path planning was implemented by looking for the shortest path within a probabilistic roadmap (PRM) [9]. Although a shortest path can be obtained by searching in the gridmap directly, with the use of PRM we account for the kinematics of the mobile platform during the construction phase. Thus, the output is a set of kinematically feasible robot poses rather than just cells.

Moreover, the marginal pose posterior was hypothesized through the Pose SLAM engine for each of the simulated trajectories, and only those trajectories with an entropy measure below a given threshold were chosen as safe exploratory routes. This effectively limits the length of exploratory paths to a cumulative open loop uncertainty threshold. Figure 5.1b shows in red all frontier regions. Of these, actions 1 and 2, frame (c), were considered safe to reach.

5.1.2 Place Re-Visiting Actions

In contrast to exploratory actions, place re-visiting actions are intended to improve localization of the robot, which translates into a decrease in entropy. In [4] a topological map is built to search for loop closures. In our case, the Pose SLAM map readily provides this topological structure. The search for loop closure candidates in our case uses the very same mechanisms of data association introduced in Pose SLAM [3], and hence, takes explicitly into account the uncertainty in localization when computing the distance between poses. The objective is to select the set of poses close to a distance d from the current pose, and to choose from these the one that maximizes information gain for the entire network.

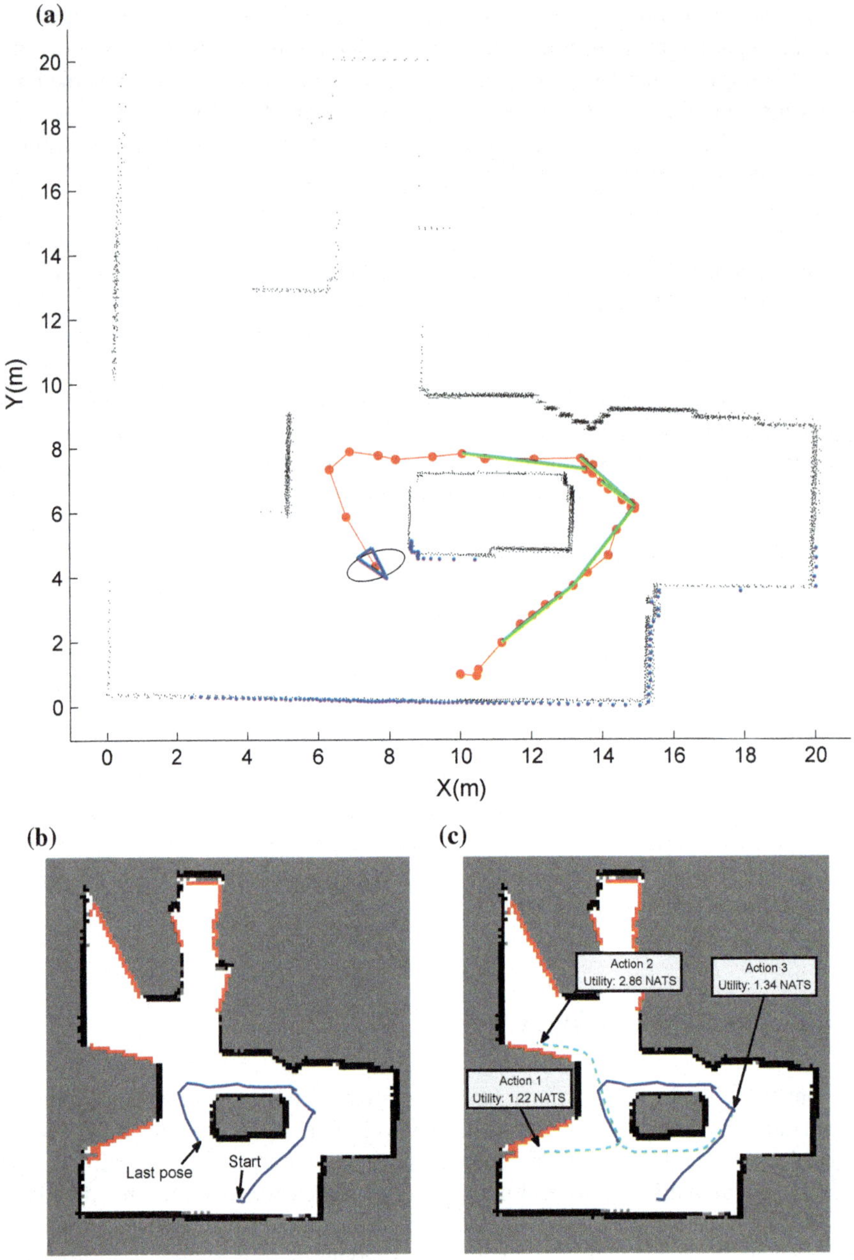

Fig. 5.1 The Pose SLAM posterior is used to render an occupancy map, which is used to generate candidate paths. **a** Pose SLAM map. **b** Gridmap and frontiers (*red cells*). **c** Candidate paths and their utilities

First, we compute a distribution of the squared distance in belief space from the current pose x_k to each other pose x_i in the map[1]

$$\mu_d = \|\mu_k - \mu_i\|^2, \tag{5.1}$$

$$\sigma_d^2 = \mathbf{H}_d \begin{bmatrix} \boldsymbol{\Sigma}_{ii} & \boldsymbol{\Sigma}_{ik} \\ \boldsymbol{\Sigma}_{ik}^\top & \boldsymbol{\Sigma}_{kk} \end{bmatrix} \mathbf{H}_d^\top . \tag{5.2}$$

We do not want to consider neither loop closures that make the robot return large paths nor those that connect only to nearby neighbors. The probability of pose x_i being at a squared distance d_r with a threshold v to pose x_k is

$$p_d = \int_{d_r - v}^{d_r + v} \mathcal{N}(\mu_d, \sigma_d^2) \; . \tag{5.3}$$

The parameter d_r sets the mean squared distance to consider and the parameter v sets the window search size. Small values indicate that we want to consider loops strictly at a squared distance d_r from the current location, whereas large values would be more permissive. This probability, for a Gaussian distribution is easily evaluated with the error function (`erf`). If the probability of being within the range $(d_r - v, d_r + v)$ (in belief space) is above a given threshold, the pose is added to the set of loop closure candidates.

Next, from this set of loop closure candidates we select the one that provides the largest information gain, computed with Eq. 3.20. Continuing with the example in Fig. 5.1c, it is action 3 the loop closure sequence that fulfills this condition. In contrast to the active loop closing technique used in [4], the technique discussed here accounts for the uncertainty in the trajectory, and is therefore more robust to localization errors. Finally, a path to the loop closure candidate is computed using the same approach as with the exploration candidates.

5.2 Utility of Actions

Once we have computed a set of candidate paths, we need to calculate their utility and select the one with largest reward. Our utility function is the expected information gain, i.e., the decrease in entropy for the posterior.

Just as in [4], we can approximate the full entropy of the trajectory and map as the sum of the individual entropies. That is, the joint entropy of a trajectory $\mathbf{x}_k = x_{1:k}$ and a map $\mathbf{O}_m$, given a set of motion commands $\mathcal{U}_k = u_{1:k}$ and a set of observations $\mathcal{Z}_k = z_{1:k}$ is

[1]With a slight abuse in notation, μ_i refers here only to the x and y components of μ_i, and $\boldsymbol{\Sigma}_{ii}$ to the marginal elements of $\boldsymbol{\Sigma}_{ii}$, leaving the orientation terms out. The Jacobian $\mathbf{H}_d$ is simply $2[(\mu_i - \mu_k)^\top, (\mu_k - \mu_i)^\top]$.

$$H(\mathbf{x}_k, \mathbf{O}_m|\mathcal{U}_k, \mathcal{Z}_k) = H(\mathbf{x}_k|\mathcal{U}_k, \mathcal{Z}_k) + \int_x p(\mathbf{x}_k|\mathcal{U}_k, \mathcal{Z}_k) H(\mathbf{O}_m|\mathbf{x}_k, \mathcal{U}_k, \mathcal{Z}_k) dx$$
$$\approx H(\mathbf{x}_k|\mathcal{U}_k, \mathcal{Z}_k) + H(\mathbf{O}_m|\mathcal{U}_k, \mathcal{Z}_k). \quad (5.4)$$

The entropy of the path in Pose SLAM, being a multivariate Gaussian, is given by

$$H(\mathbf{x}_k|\mathcal{U}_k, \mathcal{Z}_k) = \ln((2\pi e)^{(n/2)}|\boldsymbol{\Sigma}|), \quad (5.5)$$

where n is the dimension of the whole state vector.

Unfortunately, the evaluation of Eq. 5.5 has a drawback. As noted in [10], the covariance might easily become ill defined, with full correlated confidence and one or more eigenvalues near 0. This happens for instance when two poses become fully correlated, shrinking the probability distribution along a linear combination of states, while no information is gained in other dimensions.

To overcome this situation, we approximate the entropy of the trajectory without taking into account the correlation between poses, but averaging instead over the individual marginals [4]

$$H(\mathbf{x}_k|\mathcal{U}_k, \mathcal{Z}_k) \approx \frac{1}{k} \sum_{i=1}^{k} \ln((2\pi e)^{(n'/2)}|\boldsymbol{\Sigma}_{ii}|), \quad (5.6)$$

where n' is the dimension of the individual pose vector. Another option would be to use an a-optimal measure of information for the path such as the trace of $\boldsymbol{\Sigma}$ [10]. In our experiments we have experienced better results with the average form than the aggregated measure (trace) when used in combination with the map entropy as in Eq. 5.4. The main reason is that the effect of path length is averaged in the first case. This is a reasonable choice since we have already settle for a range of path lengths as discussed in Sect. 5.1.2.

For a map $\mathbf{O}_m$ with cell size w, the entropy is computed with

$$H(\mathbf{O}_m|\mathcal{U}_k, \mathcal{Z}_k) = -w^2 \sum_{c \in \mathbf{O}_m} (p(c) \ln p(c) + (1 - p(c)) \ln(1 - p(c))). \quad (5.7)$$

To compute this posterior, we must hypothesize about unknown ray casting measurements. We take the same approach as in [4], where statistics about the change in entropy are computed as a function of the number of unclassified cells covered by a hypothetical laser beam. When an unknown cell is hit, its probability contribution to Eq. 5.7 is taken from this statistic.

Fortunately, and unlike particle filters, we only have one map prior in which to simulate observations, instead of doing so for each map particle. Moreover, given that state estimation is not dependent on this map, it can be computed at a very coarse resolution, with the consequent advantages in computational cost. Another advantage of our approach in contrast to the particle filter implementation is that we do not need to arbitrarily weight the cost to reach the goal as this might bias the exploration behavior. Instead, the two techniques discussed in Sect. 5.1 guarantee that all paths in the set are of similar length (either by thresholding on open loop uncertainty during exploration, or by searching for loop closures close to a distance d from the current pose). Nonetheless, high costs in path execution mean large probabilities of becoming lost. For this reason, we enforce a replanning strategy should unforeseen loop closure occur during path execution.

Given that all actions are evaluated departing from the same prior, selecting the action or path $\mathcal{U}'$ that maximizes information gain is exactly the same as selecting the path that minimizes the entropy of the joint posterior $(\mathbf{x}', \mathbf{O}'_m)$ given the path $\mathcal{U}_k$ and upon traversing the hypothetical path $\mathcal{U}'$, and observing $\mathcal{Z}_k$ and hypothesizing about ray casted unexplored cells $\mathcal{Z}'$

$$\mathcal{U}'^* = \arg\min H(\mathbf{x}', \mathbf{O}'_m | \mathcal{U}_k + \mathcal{U}', \mathcal{Z}_k + \mathcal{Z}'). \tag{5.8}$$

For the candidate paths and utilities shown in Fig. 5.1c, actions 1 and 2 are exploratory, whereas action 3 closes a loop. Action 1 only reduces uncertainty about the environment as it drives the vehicle to an unknown region. Action 3 only reduces path uncertainty bringing the robot back to a known location. Our action selection mechanism chooses path 2, which reduces uncertainty about the environment while keeping the robot well localized.

5.3 Replanning

When planning long paths we might need to predict many observations ahead of time and, most likely, these predictions will differ substantially from the actual observations obtained when the action is executed. The most evident case is when the robot closes a large loop during path execution. The path and map estimates will change considerably and the predicted gain at the end of the path might not be relevant anymore, or even worse, the rest of the path candidate might be willing to drive the robot to a collision.

One clear alternative is to use a receding horizon to plan, but such continuous replanning is prohibitively expensive in computational terms, especially for large or finely grained maps. We opt to re-plan only when it is worth doing so. One way to know when it is wise to replan is by anticipating large deformations in the map. This occurs only if large information gains are fed to the pose network. Fortunately, these can be anticipated with the information gain in Eq. 3.20. That is, we replan if during path execution this value becomes large for any loop closure, making our predictions obsolete.

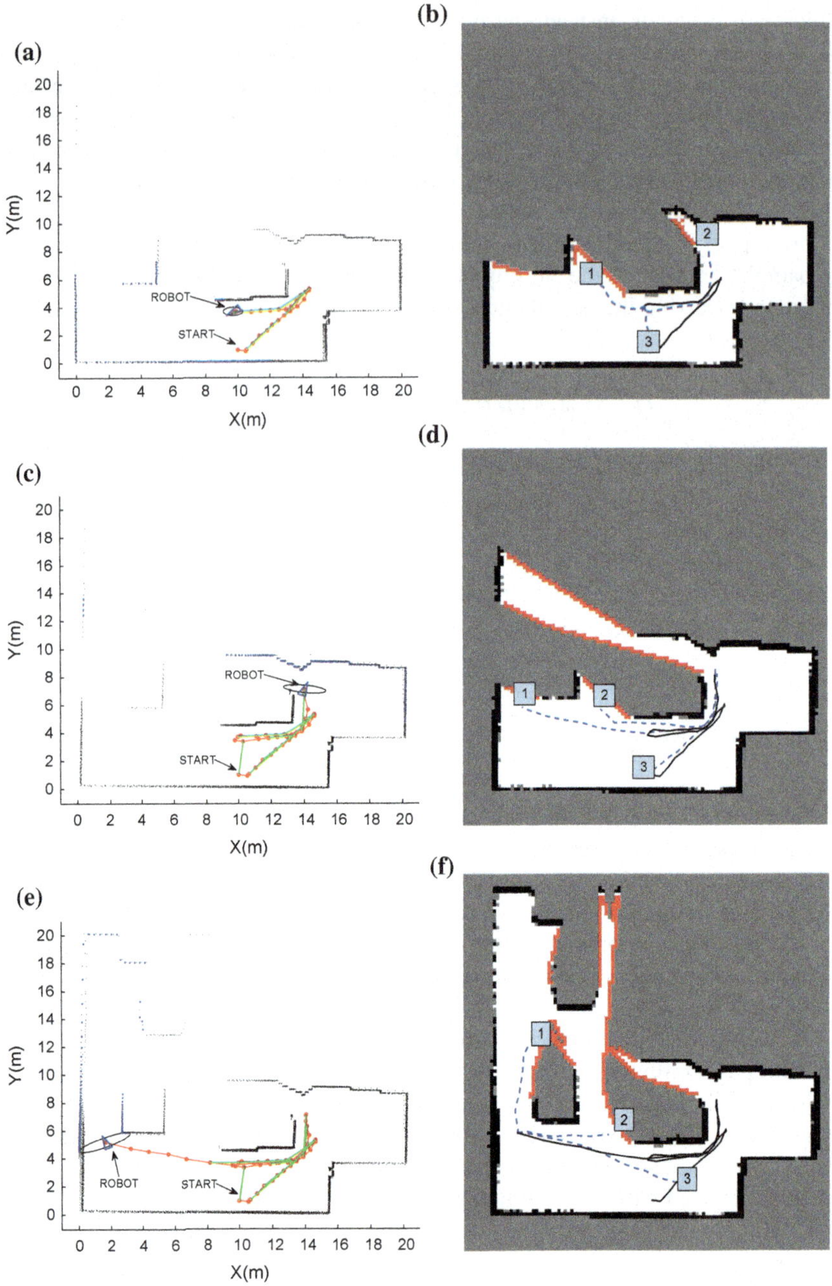

Fig. 5.2 Three points in time during the exploration process. At time step 26 (frames a and b), the robot has the following reduction in entropy: Action 1 = 1.1121 nats, Action 2 = 1.2378 nats, and Action 3 = 0.7111 nats. At time step 39 (frames c and d) Action 1 = 1.7534 nats, Action 2 = 1.4252 nats, and Action 3 = 1.1171 nats. Finally, at time step 52 (frames e and f), Action 1 = 1.8482 nats, Action 2 = 2.0334 nats, and Action 3 = 1.7042 nats. The actions chosen are 2, 1, and 2, respectively

5.4 Experiments

In order to evaluate the exploration strategy presented in this Chapter we simulated a robot exploring the widely used cave-like two-dimensional environment available from [11], scaled to a resolution of 20 m $\times$ 20 m. We present results of the evolution of the exploration method, the effects of replanning, and a comparison with frontier based exploration [7].

In the reported setting, robot motion was simulated with an odometric sensor with noise covariance $\boldsymbol{\Sigma}_u = \text{diag}(0.1\,\text{m}, 0.1\,\text{m}, 0.0026\,\text{rad})^2$. Moreover, a laser range finder sensor was simulated to establish links between any two poses closer than ± 3 m in x and y, and ± 0.52 rad in orientation. Relative motion constraints were measured using the iterative closest point algorithm. Measurement noise covariance was fixed at $\boldsymbol{\Sigma}_z = \text{diag}(0.05\,\text{m}, 0.05\,\text{m}, 0.0017\,\text{rad})^2$. Laser scans were simulated by ray casting over a ground truth gridmap of the environment using the true robot path. The initial uncertainty of the robot pose was set to $\boldsymbol{\Sigma}_0 = \text{diag}(0.1\,\text{m}, 0.1\,\text{m}, 0.09\,\text{rad})^2$. Nearby poses were detected with γ at 2.5 nats.

5.4.1 Exploration

The algorithm was executed with the aforementioned conditions and the effects of the exploration strategy were recorded. Figure 5.2 shows the obtained maps at three points in time. At each iteration, using the Pose SLAM prior (top row), a gridmap is rendered (bottom row) and used to compute the next exploration path. For instance, at time step 26 (frames a and d), the algorithm chooses Action 2, leading the robot to explore a region to reduce map entropy. Then, at time step 39, the shortest path planner does not find a path to the nearest frontier. The free cells to reach it form a narrow hallway which cannot be safely traversed. Instead, the path planner selects another frontier. Eventually, the algorithm chooses Action 1 because along this path the robot observes more unknown cells with the consequent larger reduction in map entropy. Finally, at time step 52, the more conservative Action 2 is selected this time since it reduces both the path and map entropies. Figure 5.3 shows the path and map entropy evolution for the execution of the entire exploration sequence.

5.4.2 Replanning

The exploration strategy can be improved with a replanning scheme. Replanning is triggered when we detect significant change between the entire Pose SLAM prior and posterior upon loop closure. It is an indicator of significant shift in the map estimate and the scheme is devised to anticipate those changes. In the reported experiments, replanning is triggered upon loop closure with information content greater than 4 nats.

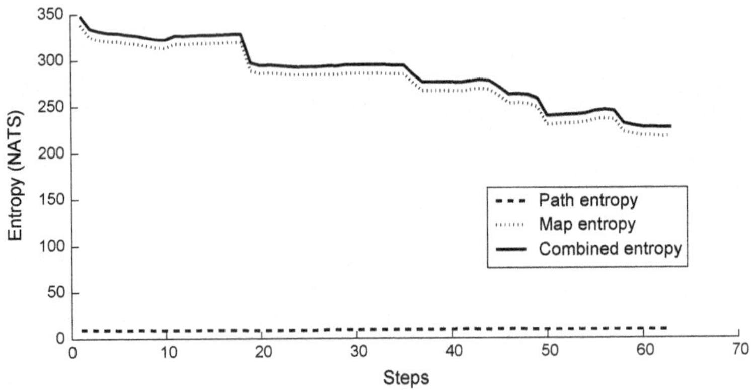

Fig. 5.3 Entropy evolution

Figure 5.4 shows a comparison of the exploration results with and without replanning. A slight drop in map entropy is observed when replanning is considered, from 147.89 nats to 146.23 nats for experimental runs of the same duration of 180 time steps. While the changes in the final map are subtle (some different regions covered, and slight improvement of entropy reduction), the changes in the localization estimates are more evident. We could also notice that the replanning strategy not only helps reduce overall map uncertainty, but also enforces better robot localization, maintaining full path entropy bounded.

5.4.3 Comparison with Frontier-Based Exploration

Next, we compare our method against pure frontier-based exploration using the same environment and specifications employed in the aforementioned experiments. Frontier-based exploration always drives the robot to the closest frontier disregarding uncertainty in the map and its localization. In our implementation analyzed frontiers are limited to a size larger than 9 cells. See Fig. 5.5. One can note that, although this greedy scheme eventually covers all the environment, the resulting map and path contain severe localization errors, as the robot barely closes three loops, which are not enough to correct the drift, causing it to end up with a final map entropy of 152.62 nats. In contrast, the Active Pose SLAM approach presented in this Chapter also covers the whole environment in the same number of time steps, yielding a slightly lower final map entropy of 146.23 nats for the same experimental setting (see Fig. 5.4c–d), thus better satisfying the competing objectives of coverage and accuracy.

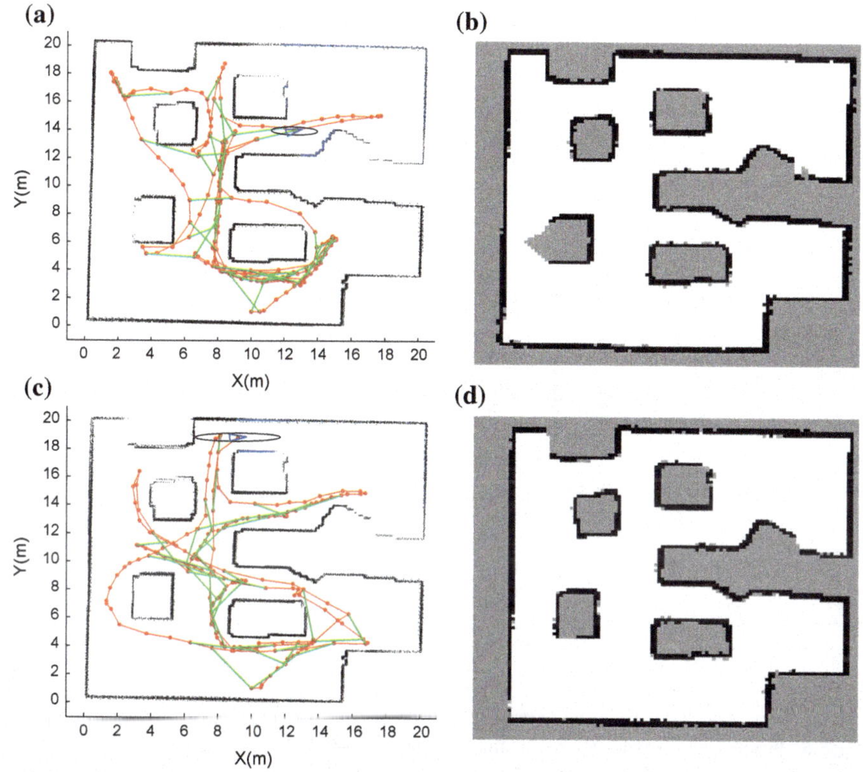

Fig. 5.4 Exploration with and without replanning. **a** Pose SLAM map and **b** gridmap made without replanning, with a final map entropy of 147.89 nats. **c** Pose SLAM map and **d** gridmap made with replanning, with a final map entropy of 146.23 nats

5.4.4 Exploration with Graph SLAM Approaches

As in our path planning method, the exploration strategy introduced in this Chapter can also be directly used even if we estimate the pose graph with a different pose-based SLAM approach. Next, we show a comparison of exploring the same environment with our exploration strategy using different SLAM techniques to maintain an estimate of the map. To this end, we employ Pose SLAM and two different Graph SLAM approaches to maintaining the map estimate. As in Sect. 4.3.6, we employ g2o[2] [12] and GTSAM[3] [13–15].

In the Graph SLAM paradigm, the problem is decoupled in two tasks: graph constructing (SLAM front-end) and graph optimization (SLAM back-end). The two Graph SLAM approaches that we employ are back-ends, i.e. they assume a graph

[2]Code accessed from: https://github.com/RainerKuemmerle/g2o.

[3]GTSAM Version 3.2.1 accessed from: https://collab.cc.gatech.edu/borg/gtsam.

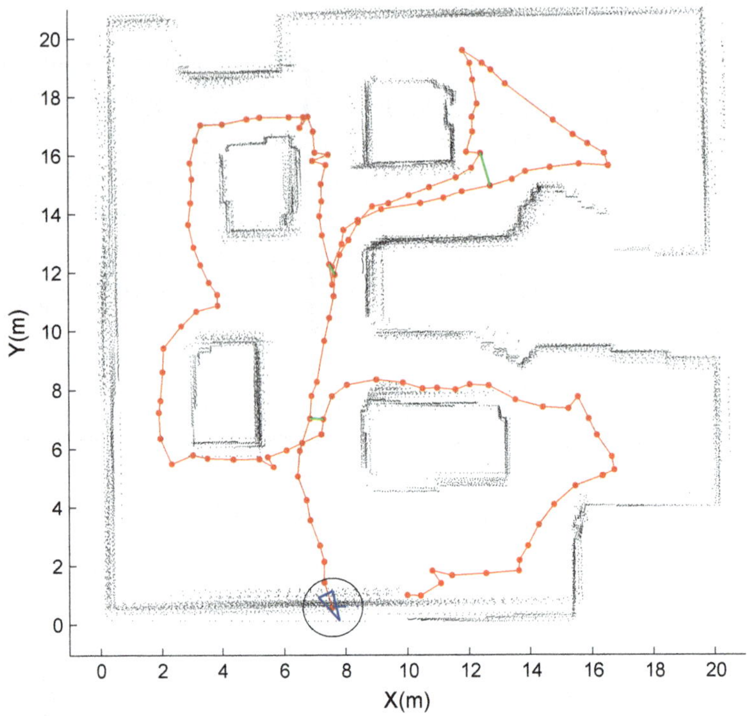

Fig. 5.5 Frontier-based exploration. The final map entropy is only reduced to 152.62 nats. Contrary to the proposed approach, this technique does not evaluate the need for loop closure and as a consequence, severe localization errors are evident at the end of the trajectory

has been built beforehand. Fortunately, Pose SLAM provides techniques to select the links and poses to build the graph. Therefore, to execute our exploration strategy we used Pose SLAM as a front-end to construct the graph, but employ the two different graph SLAM methods for its optimization.

We simulated a robot exploring the cave-like environment available in [11] and execute our exploration strategy using Pose SLAM, g2o and GTSAM to optimize the graph the robot is building autonomously. In the reported setting, robot motion was simulated with an odometric sensor with noise covariance $\boldsymbol{\Sigma}_u = \text{diag}(0.1\,\text{m}, 0.1\,\text{m}, 0.0026\,\text{rad})^2$. Moreover, a laser range finder sensor was simulated to establish links between any two poses closer than $\pm 3\,\text{m}$ in x and y, and $\pm 0.52\,\text{rad}$ in orientation. Relative motion constraints were measured using the iterative closest point algorithm. Measurement noise covariance was fixed at $\boldsymbol{\Sigma}_z = \text{diag}(0.05\,\text{m}, 0.05\,\text{m}, 0.0017\,\text{rad})^2$. Laser scans were simulated by ray casting over a ground truth gridmap of the environment using the true robot path. The initial uncertainty of the robot pose was set to $\boldsymbol{\Sigma}_0 = \text{diag}(0.1\,\text{m}, 0.1\,\text{m}, 0.09\,\text{rad})^2$. Nearby poses were detected with γ at 1 nats.

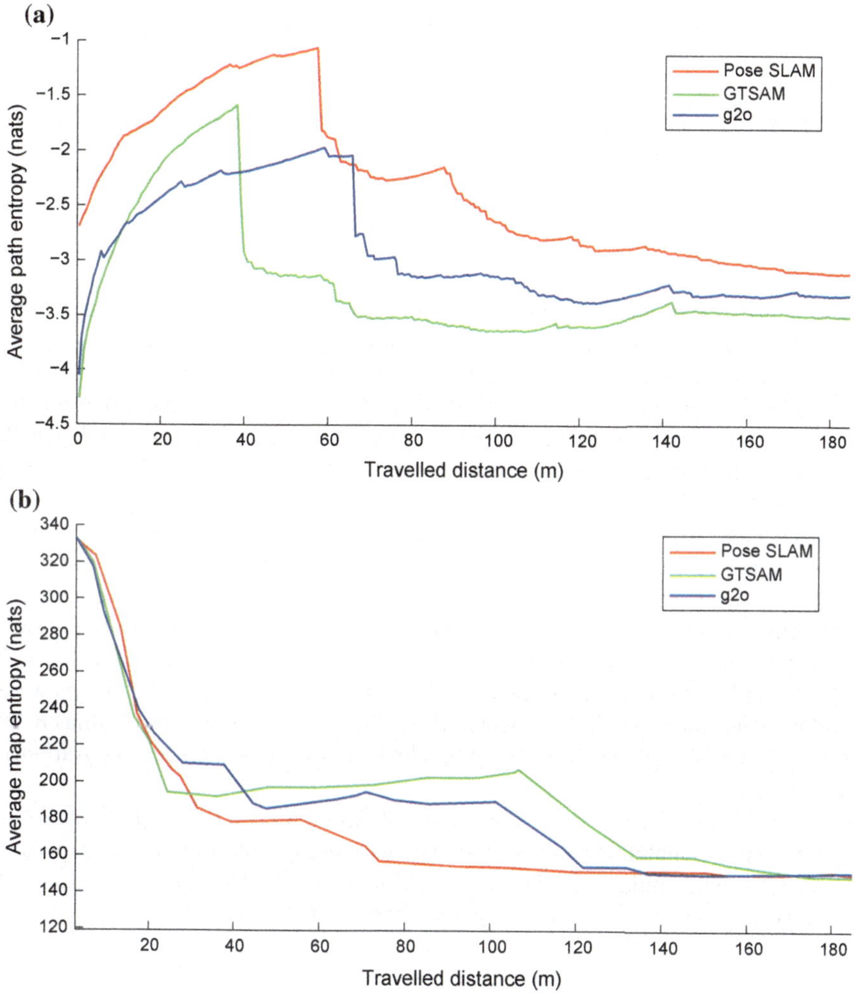

Fig. 5.6 Exploration using Pose SLAM, GTSAM and g2o to maintain the path estimate

Figure 5.6 shows the average path and map entropy evolution of 7 simulation runs of a 300 steps exploration sequence of our exploration strategy using Pose SLAM, g2o and GTSAM to maintain the map estimate. In Table 5.1, we show the average number of closed loops, coverage (measure as the number of cells labeled in the occupancy map), Absolute Trajectory Errors (ATE)[4] as well as the final path and map average entropies.

As it was discussed in Sect. 4.3.6, the values of the covariance estimates obtained with each SLAM method differ. This difference is evident in the path entropy evo-

[4]ATE implementation from the Rawseed Project (http://www.rawseeds.org) was used.

Table 5.1 Exploration using Pose SLAM, GTSAM and g2o

SLAM method	Path entopy (nats)	Map entopy (nats)	Closed loops	Coverage (m^2)	ATE (m)
Pose SLAM	−3.169	149.198	75	331.44	0.1662
GTSAM	−3.508	148.926	77	332.01	0.1649
g2o	−3.303	150.274	76	331.68	0.1663

lution plots shown in Fig. 5.6. On the other hand, we can note that the shape of the entropy evolution is similar in the three cases, which is the result of employing the same exploration strategy, which controls the growth in uncertainty in the same way. At the beginning, our exploration method let the robot explore the environment, thus increasing its localization uncertainty, but just after it reaches a path length between 60 m and 80 m the robot is driven to a region where it can close a loop to reduce its localization uncertainty.

In any case, such differences in the covariance estimate did not affect the general exploration performance obtained with the three methods. The quality of the mean estimates were similar in the three cases, as can be seen by the ATE values obtained for a path consisting of 300 robot poses. Moreover, from Fig. 5.6b, for the three cases, the average map entropy converged to a value around 148nats after exploring 180 m with similar coverage values as well. Also, the number in the closed loops were similar in the three SLAM methods, where the differences can be explained by using the same thresholds during map building and by the differences in covariance estimates.

In conclusion, since the goal of an active SLAM approach is to explore the environment while maintaining an accurate map, we can say that the exploration performance remained similar with the three SLAM methods as the trajectory errors and coverage values were pretty similar in the three cases.

5.5 Bibliographical Notes

Exploration strategies driven by uncertainty reduction date back to the seminal work of Whaite and Ferrie in [16] for the acquisition of 3-D models of objects from range data. Within the context of SLAM, it is the work of Feder et al. [17], who first proposed a metric to evaluate uncertainty reduction as the sum of the independent robot and landmark entropies with an exploration horizon of one step to autonomously produce occupancy maps.

Subsequent works that address autonomous exploration for SLAM differentiate best between each other by their coverage strategies, objective functions, and by their action selection methods. Next we show a classification based on these aspects.

5.5.1 Coverage Mechanisms

Coverage mechanisms are implemented in order to motivate exploration, that is, they avoid the robot get stuck in the same region. So as to determine coverage, it is a common practice to employ an additional map and perform frontier-based exploration [18], using an occupancy grid [4, 10, 19] or a visual map [20]. This is due to the fact that feature-based maps, which are usually computed in SLAM, do not provide straightforward ways to distinguish between known an unknown areas.

Moreover, coverage can also be introduced implicitly in the cost function. In [19–21] free parameters are used in the objective function to motivate exploration. In [22–26] the world is populated with a set of dummy landmarks, and by finding the actions that reduce the uncertainty of all landmarks and the robot pose, the coverage constraint is implicitly considered.

Alternative approaches that do not maintain an additional model apply sensor-based heuristics to cover the environment. Sensor-based heuristics extract frontiers by searching for gaps in the environment. A data structure called Gap Navigation Tree (GNT) is proposed in [27] to do this. In [28], the authors formulate a probabilistic apporach for the GNT employing a partially-complete SLAM map.

5.5.2 Objective Functions

Objective functions in Active SLAM are usually information theoretic functions, i.e. based only on information metrics. However, sometimes other constraints are also considered, which can be thought as decision-theoretic functions.

Entropy-based objective functions are used in [17, 19], where it is assumed independence between map features and the vehicle pose. Besides reducing uncertainty, Bourgault et al. [19] proposed an objective function that also motivates coverage using only information metrics.

In [10] Sim and Roy studied the information theoretic metrics for the Active SLAM problem. They pointed out that approaches whose objective functions assume independence between map features and the vehicle position will tend to underestimate the uncertainty of covariance matrices by ignoring the cross-correlations. They also provided an insight into the use of entropy as a cost function, arguing that although we can reduce faster the uncertainty of our distribution in some dimensions with such a metric, in others there is no gain of information. As an alternative, they employ the relative entropy computed as an *a-optimal* measure.

Mutual information is used in [21], where the authors perform Active SLAM with aerial vehicles. In order to control a single camera performing Bearing-only SLAM, Vidal et al. [23] also used an objective function based on mutual information. In their work, translations and orientations are kinematically decoupled, as they employ an unconstrained moving camera. They employ mutual information only to compute translations, while orientation commands are obtained using the trace of the Fisher information. Kullback–Leibler divergence is employed in [29] as part of an objective function that considers both information gain and information loss due to inconsistency in the filter.

On the other hand, while exploring, the robot is subject to other constraints such as time, path length, or energy. Overall utility functions include such constraints as well as information-theoretic metrics. These approaches are closely related to decision theory, where the goal is to choose the optimal action by maximizing an overall utility function (e.g., tradeoff between cost and benefit). However, they usually need to set free parameters to combine different units consistently in the same expression.

In [30] an exploration strategy for SLAM is proposed, where the objective function is a combination of costs and benefits according to the different robot tasks. In their work, costs include the path length and the size of the robot configuration uncertainty and benefits comprise the extraction of new features and travelling along the proposed global path. Makarenko et al. [31] employ an objective function that trades off the cost of exploring new terrain with respect to a potential reduction of uncertainty in the map and robot localization. y of the robot localization covariance is bigger than a given threshold. In [4], Stachniss et al. the utility of an action is evaluated by trading off the cost of executing it with the expected information gain.

5.5.3 Action Selection

The optimization approach used to select the actions that satisfy the objective function is another aspect in the problem of exploration for SLAM. Evaluation of the objective function is computationally expensive. Therefore, most of the strategies are greedy approaches. These methods find only the next action that optimizes their optimal criterion.

A common approach to increase scalability is to use a discrete action space. In [17] the robot was constrained to move a distance of 10 and 30 *cm* at each time step, and could only turn in increments of $22.5'$. In [19], the authors employed an action set composed of only three way points. A set of candidate destinations is evaluated in [21, 31–35]. Sim et al. [20, 36] use a greedy strategy to generate policies in the form of short trajectories. In [20], they employ a parameterized policy class using an online optimization with one-step lookahead. Their action set consisted of a discrete set over the curve parameter. Stachniss et al. [4] consider two types of generic actions: place re-visiting actions and exploration actions. Moreover, in [2, 23, 24], Vidal et al. evaluate information metrics for a uniformly distributed small set of actions carried out over a fixed amount of time, and choose the best next action from those.

Some works approach action selection as an optimization problem for a gradually identified nonlinear model. Receding Horizon strategies, a.k.a Model Predictive Control (MPC), solve an optimal control problem with fixed planning horizon but only the first action is executed. This idea has been pursued in some works [1, 25, 26, 37, 38]. Huang et al. [37] introduced a discussion about the problem of multi-step look-ahead exploration in the context of SLAM, arguing that multi-step active SLAM is possible when the current estimation error is small, the probability of observing new feature is low, and the computation capability is high.

The problem of exploration can be modelled as a Partially Observable Markov Decision Process (POMDP). A POMDP provides the most general framework for planning under uncertainty. In a POMDP, the knowledge about the agent's state is encoded as a *belief* (a probability distribution over all possible states) and the objective is to determine a *policy* giving the best action for each belief. Robot navigation is naturally modeled as a continuous problem but, unfortunately there are very few approaches able to deal with POMDPs in continuous spaces [39, 40]. Thus, the usual approach discretizes the problem and applies standard value iteration algorithms [41, 42]. Unfortunately, those algorithms can only deal with problems of low dimensionality. Point-value iteration algorithms [43, 44] somehow alleviate this problems focusing the planning to the reachable belief space. However, they are not efficient enough to be applied to large-scale navigation problems. Another inconvenient of standard POMDP algorithms is that they assume a known model of the environment.

An alternative that has been used successfully to solve large sequential decision making problems in both fully observable and partially observable domains is reinforcement learning. In particular, the policy search methods have been succesfully applied in control and robotics. Policy search methods are alternatives to value-based methods for the case of partially observable environments. The general idea behind these methods is to search for the optimal policy in the space of all possible policies by directly examining different policy parameterizations, bypassing the assignment of the value [45]. However, this technique requires gradients of the expected cost, which are not easy to compute in the context of SLAM because of the discontinuities of the measurement model.

In the work presented in [5, 38], Martinez Cantin et al. proposed a reinforcement learning approach to solve the problem of exploration for SLAM, their technique is based on the work presented in [46]. They employ a direct policy search approach [47], where the value function is approximated using Gaussian Processes (GP). The cost function adopted is the average mean square error (AMSE), which is expensive to approximate as it requires to simulate trajectories, however it is more robust to SLAM inconsistencies. Their algorithm works by first performing a Bayesian regression with a GP to map the policy parameters to estimates of the expected cost function, using previous simulations. Then, it selects points where GP predicts low expected cost or where GP variance is large, which is done using a statistical measure called infill function to know where to sample.

Similarly, Kollar et al. [6] employ a reinforcement learning approach to solve the exploration for SLAM problem. Their method begins with a geometric coverage planner that selects sensing points, which are in turn inserted in the state to be used as attractors, in the same way as it is done in [24, 25]. The cost function is defined as the sum of the squared errors and their action set consisted of discrete parameters of a cubic spline. So as to find policies, they adopted the Policy Search by Dynamic Programming (PSDP) [48] algorithm.

References

1. C. Leung, S. Huang, N. Kwok, G. Dissanayake, Planning under uncertainty using model predictive control for information gathering. Robot. Auton. Syst. **54**(11), 898–910 (2006)
2. T. Vidal-Calleja, A. Sanfeliu, J. Andrade-Cetto, Action selection for single camera SLAM. IEEE Trans. Syst. Man, Cybern. B, **40**(6), 1567–1581 (2010)
3. V. Ila, J.M. Porta, J. Andrade-Cetto, Information-based compact Pose SLAM. IEEE Trans. Robot. **26**(1), 78–93 (2010)
4. C. Stachniss, G. Grisetti, W. Burgard, Information gain-based exploration using Rao-Blackwellized particle filters, in *Robotics: Science and Systems I* (Cambridge, Jun 2005), pp. 65–72
5. R. Martinez-Cantin, N. de Freitas, E. Brochu, J. Castellanos, A. Doucet, A Bayesian exploration-exploitation approach for optimal online sensing and planning with a visually guided mobile robot. Auton. Robot. **27**(2), 93–103 (2009)
6. T. Kollar, N. Roy, Trajectory optimization using reinforcement learning for map exploration. Int. J. Robot. Res. **27**(2), 175–197 (2008)
7. B. Yamauchi, A frontier-based approach for autonomous exploration, in *IEEE International Symposium on Computational Intelligence in Robotics and Automation* (Monterrey, 1997), pp. 146–151
8. H. P. Moravec, A. Elfes. High resolution maps from wide angle sonar, in *Proceedings of IEEE International Conference on Robotics and Automation* (St. Louis, Mar 1985), pp. 116–121
9. L. Kavraki, P. Svestkaand, J.C. Latombe, M. Overmars, Probabilistic roadmaps for path planning in high dimensional configuration spaces. IEEE Trans. Robot. **12**(4), 566–580 (1996)
10. R. Sim, N. Roy, Global A-optimal robot exploration in SLAM, in *Proceedings of IEEE International Conference on Robotics and Automation* (Barcelona, Apr 2005), pp. 661–666
11. A. Howard, N. Roy, The robotics data set repository (Radish) (2003), http://radish.sourceforge.net
12. R. Kummerle, G. Grisetti, H. Strasdat, K. Konolige, W. Burgard, g2o: a general framework for graph optimization, in *Proceedings of IEEE International Conference on Robotics and Automation* (Shanghai, May 2011), pp. 3607–3613
13. F. Dellaert, M. Kaess, Square root SAM: simultaneous localization and mapping via square root information smoothing. Int. J. Robot. Res. **25**(12), 1181–1204 (2006)
14. M. Kaess, A. Ranganathan, F. Dellaert, iSAM: incremental smoothing and mapping. IEEE Trans. Robot. **24**(6), 1365–1378 (2008)
15. M. Kaess, H. Johannsson, R. Roberts, V. Ila, J. Leonard, F. Dellaert, iSAM2: incremental smoothing and mapping using the bayes tree. Int. J. Robot. Res. **31**(2), 216–235 (2012)
16. P. Whaite, F.P. Ferrie, Autonomous exploration: driven by uncertainty. IEEE Trans. Pattern Anal. Mach. Intell. **19**(3), 193–205 (1997)
17. H.J.S. Feder, J.J. Leonard, C.M. Smith, Adaptive mobile robot navigation and mapping. Int. J. Robot. Res. **18**, 650–668 (1999)
18. B. Yamauchi, Frontier-based exploration using multiple robots, in *International Conference on Autonomous Agents* (Minneapolis, 1998), pp. 47–53
19. F. Bourgault, A.A. Makarenko, S.B. Williams, B. Grocholsky, H.F. Durrant-Whyte, Information based adaptative robotic exploration, in *Proceedings of IEEE/RSJ International Conference on Intelligent Robots and Systems* (Lausanne, Oct 2002), pp. 540–545
20. R. Sim, G. Dudek, N. Roy, Online control policy optimization for minimizing map uncertainty during exploration, in *Proceedings of IEEE International Conference on Robotics and Automation* (New Orleans, Apr 2004), pp. 1758–1763
21. M. Bryson, S. Sukkarieh, An information-theoretic approach to autonomous navigation and guidance of an uninhabited aerial vehicle in unknown environments, in *Proceedings of IEEE/RSJ International Conference on Intelligent Robots and Systems* (Edmonton, Aug 2005), pp. 3770–3775
22. R. Sim, Stable exploration for bearings-only SLAM, in *Proceedings of IEEE International Conference on Robotics and Automation* (Barcelona, Apr 2005), pp. 2422–2427

23. T. Vidal-Calleja, A.J. Davison, J. Andrade-Cetto, D.W. Murray, Active control for single camera SLAM, in *Proceedings of IEEE International Conference on Robotics and Automation* (Orlando, May 2006), pp. 1930–1936
24. T. Vidal-Calleja, A. Sanfeliu, J.Andrade-Cetto, Guiding and localising in real-time a mobile robot with a monocular camera in non-flat terrains, in *Proceedings of 6th IFAC/EURON Symposium on Intelligent Autonomous Vehicles* (Toulouse, Sep 2007)
25. C. Leung, S. Huang, G. Dissanayake, Active SLAM using model predictive control and attractor based exploration, in *Proceedings of IEEE/RSJ International Conference on Intelligent Robots and Systems* (Beijing, 2006), pp. 5026–5031
26. C. Leung, S. Huang, G. Dissanayake, Active SLAM for structured environments, in *Proceedings of IEEE International Conference on Robotics and Automation* (Pasadena, 2008), pp. 1898–1903
27. B. Tovar, R. Murrieta-Cid, S. LaValle, Distance-optimal navigation in an unknown environment without sensing distances. IEEE Trans. Robot. **23**(3), 506–518 (2007)
28. L. Murphy, P. Newman, Using incomplete online metric maps for topological exploration with the gap navigation tree, in *Proceedings of IEEE International Conference on Robotics and Automation* (Pasadena, May 2008), pp. 2792–2797
29. L. Carlone, J. Du, M. K. Ng, B. Bona, M. Indri, An application of kullback-leibler divergence to Active SLAM and exploration with particle filters, in *Proceedings of IEEE/RSJ International Conference on Intelligent Robots and Systems* (Taipei, 2010), pp. 287–293
30. R. Bauer, W.D Rencken, Sonar feature based exploration, in *Proceedings of IEEE/RSJ International Conference on Intelligent Robots and Systems* (Pittsburgh, 1995), pp. 148–153
31. A. Makarenko, S.B. Williams, F. Bourgault, H.F. Durrant-Whyte, An experiment in integrated exploration, in *Proceedings of IEEE/RSJ International Conference on Intelligent Robots and Systems* (Lausanne, Oct 2002), pp. 534–539
32. C. Stachniss, W. Burgard, Mapping and exploration with mobile robots using coverage maps, in *Proceedings of IEEE/RSJ International Conference on Intelligent Robots and Systems*, vol. 1 (Las Vegas, 2003), pp. 467–472
33. P.M. Newman, M. Bosse, J.J. Leonard, Autonomous feature-based exploration, in *Proceedings of IEEE International Conference on Robotics and Automation* (Taipei, Sep 2003), pp. 1234–1240
34. R. Rocha, J. Dias, A. Carvalho, Cooperative multi-robot systems: a study of vision-based 3-D mapping using information theory. Robot. Auton. Syst. **53**, 282–311 (2005)
35. D. Joho, C. Stachniss, P. Pfaff, W. Burgard, Autonomous exploration for 3D map learning, in *Autonome Mobile Systems*, ed. by Karsten Berns, Tobias Luksch (Springer, Kaiserslautern, Oct 2007), pp. 22–28
36. R. Sim, G. Dudek, Effective exploration strategies for the construction of visual maps, in *Proceedings of IEEE/RSJ International Conference on Intelligent Robots and Systems* (Las Vegas, Oct 2003), pp. 3224–3231
37. S. Huang, N.M. Kwok, G. Dissanayake, Q.P. Ha, G. Fang, Multi-Step look-ahead trajectory planning in SLAM: Possibility and necessity, in *Proceedings of IEEE International Conference on Robotics and Automation* (Barcelona, 2005)
38. R. Martinez-Cantin, N. de Freitas, A. Doucet, J. Castellanos, Active policy learning for robot planning and exploration under uncertainty, in *Robotics: Science and Systems III* (Atlanta, Jun 2007), pp. 334–341
39. S. Thrun, Monte Carlo POMDPs, in *Advances in Neural Information Processing Systems 12* (The MIT Press, Apr 1999), pp. 1064–1070
40. J.M. Porta, N. Vlassis, M.T.J. Spaan, P. Poupart, Point-based value iteration for continuous POMDPs. J. Mach. Learn. Res. **7**, 2329–2367 (2006)
41. E.J. Sondik, The optimal control of partially observable Markov processes over the infinite horizon: discounted cost. Oper. Res. **26**, 282–304 (1978)
42. L.P. Kaelbling, M.L. Littman, A.R. Cassandra, Planning and acting in partially observable stochastic domains. Artif. Intell. **101**, 99–134 (1995)

43. J. Pineau, G. Gordon, S. Thrun, Point-based value iteration: an anytime algorithm for POMDPs, in *Proceedings of International Joint Conference on Artificial Intelligence* (Acapulco, 2003)
44. M.T.J. Spaan, N.A. Vlassis, Perseus: randomized point-based value iteration for POMDPs. J. Artif. Intell. Res. **24**, 195–220 (2005)
45. L. Peshkin, Reinforcement learning by policy search. Ph.D. thesis, Brown University, 1971
46. A.W. Moore, J. Schneider, Memory-based stochastic optimization, in *Neural Information Processing Systems* eds. by D. Touretzky, M. Mozer, M. Hasselmo (MIT Press, 1996), pp. 1066–1072
47. A.Y. Ng, M.I. Jordan, Pegasus: a policy search method for large MDPs and POMDPs, in *Conference on Uncertainty Artificial Intelligence* (Stanford, 2000), pp. 406–415
48. J.D. Bagnell, S. Kakade, A.Y. Ng, J. Schneider, Policy search by dynamic programming. Adv. Neural Inf. Process. Syst. **16**, (2003)

Chapter 6
Conclusions

The work presented in this book constitutes a step towards an integrated framework for mapping, planning and exploration for autonomous mobile robots. Along the book, the unifying point in our solution to such tasks was the use of the Pose SLAM approach as the basic state estimation machinery.

The key contributions of this book can be summarized as follows. It introduced a visual odometry technique, which yields motion estimates with the appropriate uncertainty bounds; an extension of Pose SLAM to work with poses in 6 DOF; a traversability map building approach; a path planning method that computes the most reliable path to the goal in a pose graph computed with Pose SLAM; and an autonomous exploration strategy to automate the belief roadmap building for Pose SLAM.

The book started with a discussion of the SLAM front-end in Chap. 2, where we presented our choice to generate the motion constraints to be fed to the estimation machinery. We described in more detail our visual odometry implementation presented in [1] and complemented it with a technique to model the uncertainty of the relative-motion constraints.

Although our visual odometry approach uses SIFT features to compute pose constraints other scale invariant feature points can also be used, such as the Speed Up Robust Features (SURF) [2]. These features have similar response properties to SIFTs, replacing Gaussian convolutions with Haar convolutions, and a significant reduction in computational cost.

The technique we introduced to model the covariance of the relative motion observations linearly propagates the noise covariances of the features obtained in a pair of stereo images through the visual odometry process. This approach is based on the use of the implicit theorem, which allowed us to compute the Jacobian of a function for which we do not have a closed form expression. The outcome of this method was evaluated with a Chi-square test with a two sided 95% probability region, which

R. Valencia and J. Andrade-Cetto, *Mapping, Planning and Exploration with Pose SLAM*, Springer Tracts in Advanced Robotics 119,
DOI 10.1007/978-3-319-60603-3_6

confirmed that the covariances computed with the error propagation approach were consistent with Monte Carlo realizations.

In our tests we approximated the covariance of the SIFT features using its scale. A further refinement would be to set this covariance proportional to the inverse of the Hessian matrix as it is done in [3]. Although this improves the uncertainty modeling, it also adds extra computations.

In Chap. 3 we presented our SLAM back-end, where we discussed the topics related only to the estimation tool. We presented two data association strategies for Pose SLAM that exploit the filter information. We introduced the first of them in [1] and an improved version appeared in [4]. For consistent estimates, we noted that both strategies are more efficient and less affected by perceptual aliasing.

We noted that Pose SLAM presents advantages related to scalability and precision by reducing the state size and the number of loop closures, considering only non-redundant poses and informative links. Thus, linearization effects are delayed, maintaining the filter consistent for longer sessions. A further extension to Pose SLAM would be to not only address the errors due to the approximation introduced by linearizations, but also the errors due to fact that Jacobians are evaluated at estimates and not at the exact value. It is addressed by the maximum likelihood mapping techniques such as [5, 6], also referred to as Pose Graph SLAM. Fortunately, the techniques that we introduced in Chaps. 4 and 5 can also accommodate the use of such SLAM approaches as well as any other delayed-state SLAM algorithms [7, 8].

We also discussed the implementation of Pose SLAM to work with poses in 6 DOF. We showed that a parameterization of rotations with Euler angles is a straightforward representation. However, it suffers from the problem of gimbal lock. This can be alleviated with a quaternion parameterization, which generates singular covariance matrices due to the normalized quaternion representation. This is fixed with the appropriate Jacobian projection. One possible extension to our 6 DOF Pose SLAM implementation is the use of the exponential map representation to avoid both issues. In this venue, a recent work by Hertzberg et al. [9] would be useful if an exponential map representation is chosen, where the authors showed a principled way to deal with manifolds in sensor fusion algorithms such as the information filter employed in this work.

In any case, the choice of any of these representations is better justified by the final application. For instance, a quaternion parameterization might be more useful for an aerial vehicle than for a terrestrial one. For the mapping results shown in Chap. 3, we chose the Euler parameterization to build a map of an urban environment with a four wheel skid steer mobile robot.

These mapping results consisted of a volumetric map made of aligned 3D point clouds, and a traversability map, useful for robot path planning and navigation; a much needed step usually neglected in most SLAM implementations. Traversability maps were derived by transforming the map of point clouds into a representation compatible with the robot kinematic characteristics. Using this technique was possible to compute the maps needed for the navigation of a heterogeneous fleet of service robots in urban settings [10]. Nevertheless, one issue with the traversability map is that it requires extensive post-processing and it does not exploit the uncertainty modeled by SLAM. In Chap. 4 we introduced a principled method for path planning under uncertainty that directly employs the Pose SLAM results without further post-processing the maps.

In Chap. 4 we argued that the poses of a Pose SLAM map can be readily used as nodes of a belief roadmap and thus, used for planning minimum uncertainty routes. The method presented in Chap. 4 shows improved navigation results when compared to standard path planning strategies. We showed evidence of this by presenting experiments over diverse datasets and a path execution test with one of our mobile robots in an outdoor setting.

The method includes a principled way to evaluate the cost of a path taking into account the uncertainty of traversing every edge in the map. The final path obtained is the safest among all the possible paths to the goal, increasing the chances to reach it. Three advantages of the proposed approach are that it is defined in the belief space, that it considers only the uncertainty added when moving between poses, and that it scales to large environments using approximated marginal covariances.

In our approach we assumed maximum likelihood actions and measurements, as it is usual when planning in belief space. With this, the mean estimate after a sequence of controls will lie at the mean of a node in the graph and the observation previously obtained at that position will be repeated. A further extension would be to keep this assumption to ease trajectory optimization techniques but then replan when the actual path deviates past a threshold as it is done in [11] or, moreover, relax this assumption as it is done in other planning approaches [12, 13].

The presented approach is adequate for scenarios where a robot is initially guided during map construction, but autonomous during execution. However, this mapping session need not be exhaustive as long as it traverses all routes the robot is intended to navigate in normal operations. Note however that the technique could also be used to plan routes in a partially built scenario during autonomous exploration. In [14] for instance, map optimization and pose optimization are jointly taken into account during exploration by defining a set of visibility nodes using a skeleton of the current map free zones, and planning an optimal path through these nodes. The same problem can also be addressed using frontiers instead as the driving nodes for exploration as we showed in Chap. 5.

In Chap. 5 we presented an active exploration strategy tightly integrated with Pose SLAM. The work is inspired in the action selection mechanisms reported in [15]. The approach needs only to compute map entropy at the Pose SLAM mean instead of at each map particle. Furthermore, the resolution of the gridmap used is independent of the Pose SLAM estimate, and it can be as coarse as needed. These two issues allow efficient computation of the information gain objective function used to evaluate candidate exploration paths, with the end result of a scalable solution to the problem.

The mechanism to evaluate place revisiting actions is simple and is tightly integrated within Pose SLAM. The selection of loop closure candidates takes into account path uncertainty and benefits from the marginal estimates maintained within Pose SLAM. The exploration strategy detects significant changes in the state estimate to interrupt the execution of large loop closure trajectories and triggers replanning. The end result is improved map and localization entropy reduction.

In the same way that replanning can be triggered upon unexpected loop closing with the consequent reduction of path uncertainty above 4 nats, it could be possible to trigger replanning upon unexpected significant improvement of map entropy before completing an exploratory trajectory. Take for instance frame e in Fig. 5.2. Passing near point 2 in the path to point 1 might increase map coverage significantly above than the current map prior, and hence, continuing exploration towards point 1 might not be the right thing to do, especially since odometric error accumulates during open loop traverse. To account for this, we need a way to evaluate overall map entropy at a higher frame rate, perhaps by only measuring information gain over the cells that become covered during path execution. The fact that entropy reduction in the path can be computed online and at high frame rate is thanks to the use of Pose SLAM as the estimation workhorse, but unfortunately, in Pose SLAM, the map posterior is marginalized out and needs to be computed to evaluate exploration candidates. Doing so at a high frame rate can be achieved with a more scalabe alternative to gridmaps, such as the one presented in [16].

Another refinement to our exploration strategy would be to use a more elaborated way to predict sensor measurements. When candidate paths are evaluated, we need to predict sensor observations. Posterior maps are predicted with the same approach as in [15], where statistics about the change in entropy are computed as a function of the number of unclassified cells covered by a hypothetical laser beam. For the case of robot path posterior, so as to predict the noise in relative pose measurements we can use statistics about the covariance of such measurements as a function of the linear and angular distances between poses. They can be computed using the error propagation approach described in Chap. 2, but adapted for each sensor modality. For the case of laser range finder sensors the error propagation can be done as described in [17].

A further improvement to our strategy would be to employ a submapping SLAM alternative in order to spend less time revisiting the same areas, thus only the areas that join submaps would require to be explored again. Likewise, our exploration strategy can be also extended to accommodate other tasks, such as it is done in [18]. That is, a SLAM back-end can be kept alive while the robot performs a specific task, such as interacting with people or performing a search and rescue task, besides solely improving its map and localization. It is out of the scope of this work, but opens a new direction of research.

Lastly, along the book we noted that our approaches can be perfectly suited for other SLAM formulations, such as Graph SLAM approaches. We give evidence of this by showing in Chap. 4 results on computing minimum uncertainty paths in maps built with Graph SLAM approaches as well as exploring an environment with our strategy introduced in Chap. 5, while maintaining the map estimate with different Graph SLAM methods. To this end, we employed state-of-the-art implementations such as g2o [19] and GTSAM [20–22]. In Chap. 4 we noted that the final paths are similar, they all drive the robot throughout practically the same regions, while in Chap. 5 the exploration performance remained similar with the different SLAM methods as the trajectory errors and coverage values were pretty similar in the three cases. The distinctions among our results are explained by the differences in the estimated pose marginals obtained from each SLAM approach and by using the same thresholds to detect neighboring poses in all the experiments. That is, with different covariance estimates distinct nodes were connected in the path search in our path planning method, while in our exploration strategy the differences are explained by using the same thresholds during map building for different values of covariance estimates obtained by each SLAM method. The difference in the covariance size is also evident in the final accumulated costs in both planning and exploration experiments. The main reason of obtaining such differences in marginal covariances are due to the way the graph is optimized in each SLAM method and the way that each marginal is recovered. In conclusion, we noted that our approaches can be directly applied to compute minimum uncertainty paths and to explore the environment independently of the state estimation machinery.

References

1. V. Ila, J. Andrade-Cetto, R. Valencia, A. Sanfeliu, Vision-based loop closing for delayed state robot mapping, in *Proceedings of IEEE/RSJ International Conference on Intelligent Robots and Systems* (San Diego, Nov 2007), pp. 3892–3897
2. H. Bay, T. Tuytelaars, L. Van Gool, SURF: Speeded up robust features, in *Proceedings of 9th European Conference on Computer Vision*. Lecture Notes in Computer Science, vol. 3951 (Graz, 2006), pp. 404–417
3. B. Zeisl, P. Georgel, F. Schweiger, E. Steinbach, N. Navab, Estimation of location uncertainty for scale invariant feature points, in *British Machine Vision Conference* (London, 2009), pp. 57.1–57.12
4. V. Ila, J.M. Porta, J. Andrade-Cetto, Information-based compact Pose SLAM. IEEE Trans. Robot. **26**(1), 78–93 (2010)
5. E. Olson, J. Leonard, S. Teller, Fast iterative alignment of pose graphs with poor initial estimates, in *Proceedings of IEEE International Conference on Robotics and Automation* (Orlando, May 2006), pp. 2262–2269
6. G. Grisetti, C. Stachniss, S. Grzonka, W. Burgard, A tree parameterization for efficiently computing maximum likelihood maps using gradient descent, in *Robotics: Science and Systems III* (Atlanta, Jun. 2007), pp. 9:1–9:8
7. R.M. Eustice, H. Singh, J.J. Leonard, Exactly sparse delayed-state filters for view-based SLAM. IEEE Trans. Robot. **22**(6), 1100–1114 (2006)
8. K. Konolige, M. Agrawal, FrameSLAM: from bundle adjustment to realtime visual mapping. IEEE Trans. Robot. **24**(5), 1066–1077 (2008)
9. C. Hertzberg, R. Wagner, U. Frese, L. Schröder, Integrating generic sensor fusion algorithms with sound state representations through encapsulation of manifolds. Inf. Fusion **14**(1), 57–77 (2013)
10. A. Sanfeliu J. Andrade-Cetto, Ubiquitous networking robotics in urban settings, in *Proceedings of IEEE/RSJ IROS Workshop on Network Robot Systems* (Beijing, Oct 2006), pp. 14–18
11. R. Platt, R. Tedrake, L. Kaelbling, T. Lozano-Perez, Belief space planning assuming maximum likelihood observations, in *Robotics: Science and Systems VI* (Zaragoza, Spain, Jun 2010)
12. A. Bry, N. Roy, Rapidly-exploring Random Belief Trees for motion planning under uncertainty, in *Proceedings of IEEE International Conference on Robotics and Automation* (Shanghai, May 2011), pp. 723–730
13. J. van den Berg, P. Abbeel, K. Goldberg, LQG-MP: optimized path planning for robots with motion uncertainty and imperfect state information, in *Robotics: Science and Systems VI*, (Zaragoza, Spain, Jun 2010)

14. T. Kollar, N. Roy, Efficient optimization of information-theoretic exploration in SLAM, in *Proceedings of 23th AAAI Conference on Artificial Intelligence* (Chicago, Jul 2008), pp. 1369–1375
15. C. Stachniss, G. Grisetti, W. Burgard, Information gain-based exploration using Rao-Blackwellized particle filters, in *Robotics: Science and Systems I* (Cambridge, Jun 2005), pp. 65–72
16. S.T. O'Callaghan, F. Ramos, Gaussian process occupancy maps. Int. J. Robot. Res. **31**(1), 42–62 (2012)
17. A. Censi, An accurate closed-form estimate of ICP's covariance, in *Proceedings of IEEE International Conference on Robotics and Automation* (Rome, Apr 2007), pp. 3167–3172
18. A. Guez, J. Pineau, Multi-tasking SLAM, in *Proceedings of IEEE International Conference on Robotics and Automation* (Anchorage, May 2010), pp. 377–384
19. R. Kummerle, G. Grisetti, H. Strasdat, K. Konolige, W. Burgard, g2o: A general framework for graph optimization, in *Proceedings of IEEE International Conference on Robotics and Automation* (Shanghai, May 2011), pp. 3607–3613
20. F. Dellaert, M. Kaess, Square root SAM: Simultaneous localization and mapping via square root information smoothing. Int. J. Robot. Res. **25**(12), 1181–1204 (2006)
21. M. Kaess, A. Ranganathan, F. Dellaert, iSAM: Incremental smoothing and mapping. IEEE Trans. Robot. **24**(6), 1365–1378 (2008)
22. M. Kaess, H. Johannsson, R. Roberts, V. Ila, J. Leonard, F. Dellaert, iSAM2: Incremental smoothing and mapping using the bayes tree. Int. J. Robot. Res. **31**(2), 216–235 (2012)

CPSIA information can be obtained
at www.ICGtesting.com
Printed in the USA
BVOW07*1227050717
488497BV00017B/2/P

9 783319 606026